Applied Probability and Statistics

BAILEY · The Elements of Stochastic Processes with Applications to the Natural Sciences

BAILEY · Mathematics, Statistics and Systems for Health

BARTHOLOMEW · Stochastic Models for Social Processes, *Second Edition*

BECK and ARNOLD · Parameter Estimation in Engineering and Science

BENNETT and FRANKLIN · Statistical Anal~ ˈ and the Chemical Industry

BHAT · Elements of Applied St~

BLOOMFIELD · Fourier Ar ⸱n

BOX · R. A. Fisher, The Life

BOX and DRAPER · Evolutio~ ⸱⸱⸱od for Process Improvement

BOX, HUNTER, and HUNTER ⸱xperimenters: An Introduction to Design, Data A ⸱⸱⸱d Model Building

BROWN and HOLLANDER · Statistics: A Biomedical Introduction

BROWNLEE · Statistical Theory and Methodology in Science and Engineering, *Second Edition*

BURY · Statistical Models in Applied Science

CHAMBERS · Computational Methods for Data Analysis

CHATTERJEE and PRICE · Regression Analysis by Example

CHERNOFF and MOSES · Elementary Decision Theory

CHOW · Analysis and Control of Dynamic Economic Systems

CLELLAND, deCANI, BROWN, BURSK, and MURRAY · Basic Statistics with Business Applications, *Second Edition*

COCHRAN · Sampling Techniques, *Third Edition*

COCHRAN and COX · Experimental Designs, *Second Edition*

COX · Planning of Experiments

COX and MILLER · The Theory of Stochastic Processes, *Second Edition*

DANIEL · Application of Statistics to Industrial Experimentation

DANIEL · Biostatistics: A Foundation for Analysis in the Health Sciences, *Second Edition*

DANIEL and WOOD · Fitting Equations to Data

DAVID · Order Statistics

DEMING · Sample Design in Business Research

DODGE and ROMIG · Sampling Inspection Tables, *Second Edition*

DRAPER and SMITH · Applied Regression Analysis

DUNN · Basic Statistics: A Primer for the Biomedical Sciences, *Second Edition*

DUNN and CLARK · Applied Statistics: Analysis of Variance and Regression

ELANDT-JOHNSON · Probability Models and Statistical Methods in Genetics

FLEISS · Statistical Methods for Rates and Proportions

GALAMBOS · The Asymptotic Theory of Extreme Order Statistics

GIBBONS, OLKIN, and SOBEL · Selecting and Ordering Populations: A New Statistical Methodology

GNANADESIKAN · Methods for Statistical Data Analysis of Multivariate Observations

GOLDBERGER · Econometric Theory

GOLDSTEIN and DILLON · Discrete Discriminant Analysis

continued on back

The elements of
Stochastic Processes

A WILEY PUBLICATION IN APPLIED STATISTICS

The elements of Stochastic Processes

with applications to the natural sciences

NORMAN T. J. BAILEY

Reader in Biometry
University of Oxford

John Wiley & Sons, Inc.
New York · London · Sydney

ISBN 0 471 04165 3

Library of Congress Catalog Card Number: 63-23220

Printed in the United States of America

πάντα ῥεῖ, οὐδὲν μένει.

(Everything flows, nothing is stationary)

HERACLITUS

"... I love everything that flows, everything that has time in it and becoming, that brings us back to the beginning where there is never end: ..."

HENRY MILLER, *Tropic of Cancer*

v

Preface

Stochastic models are being used to an ever-increasing extent by those who wish to investigate phenomena that are essentially concerned with a flow of events in time, especially those exhibiting such highly variable characteristics as birth, death, transformation, evolution, etc. The rigorous probability theory of these processes presents very considerable mathematical difficulties, but no attempt has been made to handle this aspect here. Instead, the more heuristic approach of applied mathematics has been adopted to introduce the reader to a wide variety of theoretical principles and applied techniques, which are developed simultaneously.

An acquaintance with basic probability and statistics is assumed. Some knowledge of matrix algebra is required, and complex variable theory is extensively used in the latter half of the book. Differential equations, both ordinary and partial, occur frequently, though none is of a very advanced nature. Where special mathematical methods are required repeatedly, such as the use of generating functions or the solution of linear partial differential equations, additional sections are included to give a detailed discussion.

The book starts with a general introduction, and then a whole chapter on generating functions. The next four chapters deal with various kinds of processes in discrete time such as Markov chains and random walks. This is followed by the treatment of continuous-time processes, including such special applications as birth-and-death processes, queues, epidemics, etc. After this there is a discussion of diffusion processes. The two final chapters deal with the use of various approximate methods of handling certain kinds of processes, and a brief introduction to the treatment of some non-Markovian processes.

It is hoped that students will find the book useful as a first text in the theory and application of stochastic processes, after which they will be better equipped to tackle some of the more mathematically sophisticated treatises. Readers who already have some acquaintance with the subject may find the discussions of special topics to be of value.

The text of this book is based on a course of lectures given during the fall and winter quarters, 1961-2, at the Department of Biostatistics, The Johns Hopkins University, Baltimore. It gives me great pleasure to acknowledge the encouragement received from Professor Allyn Kimball, while I was a Visiting Professor in his department. I am also indebted to the Milbank Memorial Fund of New York City for support during my tenure at The Johns Hopkins University. A shortened version of the lectures was presented at the same time at the National Institutes of Health, Bethesda, where I was fortunate to receive further stimulus from Dr. Jerome Cornfield and his colleagues.

The first six chapters were drafted on Sanibel Island, Florida, during February 1962. The following nine chapters were completed during my appointment as Visiting Professor in the Department of Statistics, Stanford University, in the spring and summer quarters of 1962. I should specially like to thank Professor Herbert Solomon for arranging the financial support for the production and distribution of the multilith draft of the first fifteen chapters. I am also greatly indebted to Mrs. Betty Jo Prine, who supervised the latter, and to Mrs. Carolyn Knutsen, who undertook the actual typing, a task which she performed with great rapidity, enthusiasm, and accuracy.

The final chapter was completed in Oxford, England, during the fall of 1962, and the usual introductory material, problems for solution, references, etc., were added. A revized version of the whole book was then prepared, and I am delighted to be able to acknowledge the help received from Professor Ralph Bradley, who read most of the draft typescript and provided me with detailed comments and criticisms. I must add, of course, that any errors that remain are entirely my own responsibility. I should also like to thank Dr. J. F. C. Kingman of the University of Cambridge for making a number of helpful comments and suggestions.

Finally, I must thank my secretary in Oxford, Mrs. Kay Earnshaw, for her assistance in preparing the final version of the book for printing.

NORMAN T. J. BAILEY

Oxford
October, 1963

Contents

CHAPTER   PAGE

1   INTRODUCTION AND GENERAL ORIENTATION   1

2   GENERATING FUNCTIONS   5
   2.1   Introduction   5
   2.2   Definitions and elementary results   5
   2.3   Convolutions   7
   2.4   Compound distributions   9
   2.5   Partial fraction expansions   10
   2.6   Moment- and cumulant-generating functions .   .   .   12
      Problems for solution   14

3   RECURRENT EVENTS   16
   3.1   Introduction   16
   3.2   Definitions   16
   3.3   Basic theorems   18
   3.4   Illustration   19
   3.5   Delayed recurrent events   21
      Problems for solution   22

4   RANDOM WALK MODELS   24
   4.1   Introduction   24
   4.2   Gambler's ruin   25
   4.3   Probability distribution of ruin at nth trial .   .   .   28
   4.4   Extensions   34
      Problems for solution   36

5   MARKOV CHAINS   38
   5.1   Introduction   38
   5.2   Notation and definitions   39
   5.3   Classification of states   41
   5.4   Classification of chains   46
   5.5   Evaluation of $\mathbf{P}^n$ 47
   5.6   Illustrations   51
      Problems for solution   56

6 DISCRETE BRANCHING PROCESSES 58
  6.1 Introduction 58
  6.2 Basic theory 59
  6.3 Illustration 63
      Problems for solution 64

7 MARKOV PROCESSES IN CONTINUOUS TIME 66
  7.1 Introduction 66
  7.2 The Poisson process 67
  7.3 Use of generating functions 69
  7.4 "Random-variable" technique 70
  7.5 Solution of linear partial differential equations . . 74
  7.6 General theory 75
      Problems for solution 82

8 HOMOGENEOUS BIRTH AND DEATH PROCESSES 84
  8.1 Introduction 84
  8.2 The simple birth process 84
  8.3 The general birth process 88
  8.4 Divergent birth processes 89
  8.5 The simple death process 90
  8.6 The simple birth-and-death process 91
  8.7 The effect of immigration 97
  8.8 The general birth-and-death process 101
  8.9 Multiplicative processes 102
      Problems for solution 105

9 SOME NON-HOMOGENEOUS PROCESSES 107
  9.1 Introduction 107
  9.2 The Pólya process 107
  9.3 A simple non-homogeneous birth-and-death process . 110
  9.4 The effect of immigration 115
      Problems for solution 116

10 MULTI-DIMENSIONAL PROCESSES 117
  10.1 Introduction 117
  10.2 Population growth with two sexes 119
  10.3 The cumulative population 120
  10.4 Mutation in bacteria 125
  10.5 A multiple-phase birth process 131
      Problems for solution. 135

11 QUEUEING PROCESSES 136
  11.1 Introduction 136
  11.2 Equilibrium theory 137
  11.3 Queues with many servers 143

11.4 Monte Carlo methods in appointment systems . . . 147
11.5 Non-equilibrium treatment of a simple queue . . . 149
11.6 First passage times 157
    Problems for solution 160

12 EPIDEMIC PROCESSES 162
12.1 Introduction 162
12.2 Simple epidemics 164
12.3 General epidemics 169
12.4 Recurrent epidemics 177
12.5 Chain-binomial models 182
    Problems for solution 184

13 COMPETITION AND PREDATION 186
13.1 Introduction 186
13.2 Competition between two species 187
13.3 A prey–predator model 189

14 DIFFUSION PROCESSES 194
14.1 Introduction 194
14.2 Diffusion limit of a random walk 195
14.3 Diffusion limit of a discrete branching process . . . 197
14.4 General theory 199
14.5 Application to population growth 205

15 APPROXIMATIONS TO STOCHASTIC PROCESSES 207
15.1 Introduction 207
15.2 Continuous approximations to discrete processes . . . 207
15.3 Saddle-point approximations 214
15.4 Neglect of high-order cumulants 217
15.5 Stochastic linearization 219

16 SOME NON-MARKOVIAN PROCESSES 222
16.1 Introduction 222
16.2 Renewal theory and chromosome mapping . . . 224
16.3 Use of integral equations 228

REFERENCES 234
SOLUTIONS TO PROBLEMS 237
AUTHOR INDEX 243
SUBJECT INDEX 245

CHAPTER 1

Introduction and General Orientation

The universal spectacle of birth and death, growth and decay, change and transformation has fascinated mankind since the earliest times. Poets and philosophers alike have been preoccupied with the remorseless flow of events. In ancient Greece Heraclitus made the idea of perpetual change and flow the central concept of his philosophy. Sometimes the whole process of change has been conceived as an endless series of repetitive cycles and sometimes as a continuous line of advance. From the time of Darwin, however, previous philosophical speculations about the possibility of continuous evolutionary development were given a sound basis in observed facts.

Since the work of Newton in the 17th Century a mathematical understanding of the physical processes of the universe has been continually broadened and deepened. Biological processes, however, have proved more difficult to master. This is largely due to the much greater inherent variability of biological material. In many, though certainly not all, areas of physics variation arises mainly from mere errors of observation, and can be averaged out to negligible proportions by the use of repeated measurements. But in biological contexts variability usually has to be accepted as basic and handled as such. Hence the tremendous emphasis now placed on the use of statistical methods in designing and analyzing experiments involving any kind of biological material.

Nevertheless, when dealing with general mathematical descriptions of biological phenomena (as opposed to the problem of interpreting critical experiments) the first step has often been to ignore substantial amounts of variation. Thus the simplest procedure in dealing with population growth is to adopt such concepts as birth-rate, death-rate, immigration-rate, etc., and to treat these as operating continuously and steadily. So that, given the rates, we can write down a differential equation whose solution specifies exactly the population size (regarded as a continuous

1

variable) at any instant of time. A good deal of actuarial work is of this kind. Again, the first mathematical accounts of the spread of epidemic disease (see Bailey, 1957, for references) assumed that in a short interval of time the number of new cases would be precisely proportional to the product of the time interval, the number of susceptibles and the number of infectious persons. Assumptions of this type lead to differential equations whose solutions predict the exact numbers of susceptibles and infectives to be found at any given time.

If we are dealing with large populations, it may be legitimate to assume that the statistical fluctuations are small enough to be ignored. In this case a *deterministic model*, i.e. the type just considered, may be a sufficiently close approximation to reality for certain purposes.

There is nothing inappropriate in predicting that the present population of, say, 51,226 persons in a small town will have grown to 60,863 in 10 years' time. Provided the latter estimate is correct to within a few hundred units, it may well be a useful figure for various economic and sociological purposes. (A finer analysis would of course take the age and sex structure into account.) But if, in a small family of four children, one of whom develops measles today, we predict the total number of cases in a week's time as 2.37, it is most unlikely that this figure will have any particular significance: a practical knowledge of the variability normally observed in epidemic situations suggests that the deterministic type of model is inappropriate here.

It is clear that the only satisfactory type of prediction about the future course of an epidemic in a small family must be on a probability basis. That is, in the family of 4 mentioned above we should want to be able to specify the *probability distribution* at any time of the existing number of cases (0, 1, 2, 3, or 4). We might also want to go further and make probability statements about the actual times of occurrence of the various cases, and to distinguish between susceptibles, infectives and those who were isolated or recovered. A model which specified the complete joint probability distribution of the numbers of different kinds of individuals at each point of time would be a *stochastic model*, and the whole process, conceived as a continuous development in time would be called a *stochastic process* (or *probability process*).

Whenever the group of individuals under consideration is sufficiently small for chance fluctuations to be appreciable, it is very likely that a stochastic representation will be essential. And it often happens that the properties of a stochastic model are markedly different from those of the corresponding deterministic analog. In short, we cannot assume that the former is merely a more detailed version of the latter. This is especially the case with queues and epidemics, for example.

There is now a very large literature dealing with stochastic processes of many different kinds. Much of it makes difficult reading, especially for those who are primarily concerned with applications, rather than with abstract formulations and fundamental theory. For the latter, the reader should consult the treatise of Doob (1953). However, it is quite possible to gain a working knowledge of the subject, so far as handling a variety of practically useful stochastic processes is concerned, by adopting the less rigorous and more heuristic approach characteristic of applied mathematics. An attempt has been made to develop this approach in the present volume.

Those readers who require an introductory course to stochastic processes should read the whole book through in the order in which it is written, but those who already possess some knowledge of the subject may of course take any chapter on its own. No special effort has been made to separate theory and applications. Indeed, this has on the contrary been deliberately avoided, and the actual balance achieved varies from chapter to chapter in accordance with what seems to the author a natural development of the subject. Thus in dealing with generating functions, the basic mathematical material is introduced as a tool to be used in later chapters. Whereas the discussion of birth-and-death processes is developed from the start in a frankly biological context. In this way a close connection is maintained between applied problems and the theory required to handle them.

An acquaintance with the elements of probability theory and basic statistics is assumed. For an extended coverage of such topics, including proofs of many results quoted and used in the present book, the reader should refer to the textbooks of Feller (1957) and Parzen (1960).

Some knowledge of matrix algebra is required (e.g. Frazer, Duncan, and Collar, 1946) though special topics such as the spectral resolution of a matrix are explained in the text.

Considerable use is made of complex variable theory, and the reader should be familiar with the basic ideas at least as far as contour integration and the calculus of residues. Many standard textbooks, such as Copson (1948), are available for this purpose. Frequent use is also made of the Laplace transform technique, though no abstruse properties are involved. In this connection some readers may like to refer to McLachlan (1953), which deals both with complex variable theory and with practical applications of Laplace transforms in a straightforward and comparatively non-rigorous fashion.

Ordinary differential equations and partial differential equations are a fairly common occurrence in this book. No advanced methods are employed, but a number of special procedures are described and illustrated in detail. At the same time the reader might find it helpful to consult some textbook such as Forsyth (1929) for a fuller exposition of standard methods of solution.

Finally, the whole subject of approximation and asymptotics is one which no applied mathematician can afford to neglect. This aspect will crop up continually throughout the book as many of the simplest stochastic models entail considerable mathematical difficulty: even moderate degrees of reality in the model may result in highly intractable mathematics! A variety of approximation techniques will therefore appear in the subsequent discussions, and it cannot be too strongly emphasized that some facility in handling this aspect is vital to making any real progress with applied problems in stochastic processes. While some space in the present book is devoted to specialized topics such as the methods of steepest descents, which is not as widely understood as it should be, the reader might like to refer to the recent book by DeBruijn (1958) for a detailed and connected account of asymptotic theory. Unfortunately, there appears to be no single book dealing with the very wide range of approximate methods which are in principle available to the applied mathematician.

It will be seen from the foregoing that the whole philosophy of the present volume is essentially rooted in real phenomena: in the attempt to deepen insight into the real world by developing adequate mathematical descriptions of processes involving a flow of events in time, and exhibiting such characteristics as birth, death, growth, transformation etc. The majority of biological and medical applications at present appear to entail processes of this type. These may be loosely termed *evolutionary*, in contradistinction to the *stationary* type of time-series process. There is a considerable literature on the latter subject alone, and in the present book we have confined our attention to the evolutionary processes. This is partly because stationary processes are adequately dealt with elsewhere; and partly because the evolutionary models are more appropriate for most biological situations. For a more rigorous outline of the mathematical theory of stochastic processes the reader may like to consult Takács (1960). Proofs of theorems are, however, generally omitted, but the book contains an excellent collection of problems with solutions. A more detailed treatment of this material is given in Parzen (1962), and for more advanced reading the now classic text by Bartlett (1955) is to be recommended. The recent book by Bharucha-Reid (1960) also contains some interesting discussions and a very extensive set of bibliographies. For a wider coverage, including time-series, Rosenblatt (1962) should be consulted.

CHAPTER 2

Generating Functions

2.1 Introduction

Before embarking on a discussion of some of the more elementary
kinds of stochastic processes, namely the *recurrent events* of Chapter 3,
it will be convenient to present the main properties of *generating functions*.
This topic is of central importance in the handling of stochastic processes
involving integral-valued random variables, not only in theoretical
analyses but also in practical applications. Moreover, all processes dealing
with populations of individuals, whether these are biological organisms,
radioactive atoms, or telephone calls, are basically of this type. Considerable use will be made of the generating function method in the sequel,
the main advantage lying in the fact that we use a single function to
represent a whole collection of individual items. And as this technique is
often thought to constitute an additional basic difficulty in studying
stochastic processes, although it frequently makes possible very substantial and convenient simplifications, it seems worth while devoting
space to a specific review of the main aspects of the subject. This chapter
is in no sense a complete treatment: for a more extensive discussion see
Feller (1957, Chapter 11 and following).

2.2 Definitions and elementary results

First, let us suppose that we have a sequence of real numbers $a_0, a_1,$
$a_2, \cdots$. Then, introducing the dummy variable x, we may define a
function

$$A(x) \equiv a_0 + a_1 x + a_2 x^2 + \cdots = \sum_{j=0}^{\infty} a_j x^j. \qquad (2.1)$$

If the series converges in some real interval $-x_0 < x < x_0$, the function

5

$A(x)$ is called the *generating function* of the sequence $\{a_j\}$. This may be regarded as a transformation carrying the sequence $\{a_j\}$ into the function $A(x)$. For the moment we can take x as real, but in more advanced applications it is convenient to work with a complex variable z.

It is clear that if the sequence $\{a_j\}$ is bounded, then a comparison with the geometric series shows that $A(x)$ converges at least for $|x| < 1$.

In many, but by no means all, cases of interest to us the a_j are probabilities, that is we introduce the restriction

$$a_j \geqslant 0, \qquad \sum_{j=0}^{\infty} a_j = 1. \tag{2.2}$$

The corresponding function $A(x)$ is then a *probability-generating function*.

Let us consider specifically the probability distribution given by

$$P\{X = j\} = p_j, \tag{2.3}$$

where X is an integral-valued random variable assuming the values $0, 1, 2, \cdots$. We can also define the "tail" probabilities

$$P\{X > j\} = q_j. \tag{2.4}$$

The usual distribution function is thus

$$P\{X \leqslant j\} = 1 - q_j. \tag{2.5}$$

We now have the probability-generating function

$$P(x) = \sum_{j=0} p_j x^j = E(x^j), \tag{2.6}$$

where the operator E indicates an expectation. We can also define a generating function for the "tail" probabilities, i.e.

$$Q(x) = \sum_{j=0} q_j x^j. \tag{2.7}$$

Note that $Q(x)$ is not a probability-generating function as defined above. Although the coefficients are probabilities, they do not in general constitute a probability distribution.

Now the probabilities p_j sum to unity. Therefore $P(1) = 1$, and

$$|P(x)| \leqslant \sum |p_j x^j|$$
$$\leqslant \sum p_j, \quad \text{if} \quad |x| \leqslant 1,$$
$$\leqslant 1.$$

Thus $P(x)$ is absolutely convergent at least for $|x| \leqslant 1$. So far as $Q(x)$ is concerned, all coefficients are less than unity, and so $Q(x)$ converges absolutely at least in the open interval $|x| < 1$.

A useful result connecting $P(x)$ and $Q(x)$ is that

$$(1 - x)Q(x) = 1 - P(x), \qquad (2.8)$$

as is easily verified by comparing coefficients on both sides.

Simple formulas are available giving the mean and variance of the probability distribution p_j in terms of particular values of the generating functions and their derivatives. Thus the mean is

$$m \equiv E(X) = \sum_{j=0} jp_j = P'(1), \qquad (2.9)$$

$$= \sum_{j=0} q_j = Q(1), \qquad (2.10)$$

where the prime in (2.9) indicates differentiation. We also have

$$E\{X(X - 1)\} = \sum j(j - 1)p_j = P''(1) = 2Q'(1).$$

Hence the variance is

$$\sigma^2 \equiv \text{var}(X) = P''(1) + P'(1) - \{P'(1)\}^2, \qquad (2.11)$$

$$= 2Q'(1) + Q(1) - \{Q(1)\}^2. \qquad (2.12)$$

Similarly we can obtain the rth factorial moment $\mu'_{[r]}$ about the origin as

$$E\{X(X - 1) \cdots (X - r + 1)\} = \sum j(j - 1) \cdots (j - r + 1)p_j$$

$$= P^{(r)}(1) \equiv rQ^{(r-1)}(1), \qquad (2.13)$$

i.e. differentiating $P(x)$ r times and putting $x = 1$.

Several other sets of quantities characterizing probability distributions, such as moments, factorial moments, and cumulants, can also be handled by means of the appropriate generating functions. For an extended discussion of the relevant theory a good standard textbook of mathematical statistics should be consulted. Some of the main results required in the present book are summarized in Section 2.6 below.

2.3 Convolutions

Let us now consider two non-negative independent integral-valued random variables X and Y, having the probability distributions

$$P\{X = j\} = a_j, \quad P\{Y = k\} = b_k. \qquad (2.14)$$

The probability of the event $(X = j, Y = k)$ is therefore $a_j b_k$. Suppose we now form a new random variable

$$S = X + Y.$$

Then the event $(S = r)$ comprises the mutually exclusive events

$$(X = 0, Y = r), (X = 1, Y = r - 1), \cdots, (X = r, Y = 0).$$

If the distribution of S is given by

$$P\{S = r\} = c_r, \tag{2.15}$$

it follows that

$$c_r = a_0 b_r + a_1 b_{r-1} + \cdots + a_r b_0. \tag{2.16}$$

This method of compounding two sequences of numbers (which need not necessarily be probabilities) is called a *convolution*. We shall use the notation

$$\{c_j\} = \{a_j\}*\{b_j\}. \tag{2.17}$$

Let us define the generating functions

$$\left. \begin{aligned} A(x) &= \sum_{j=0} a_j x^j \\ B(x) &= \sum_{j=0} b_j x^j \\ C(x) &= \sum_{j=0} c_j x^j \end{aligned} \right\}. \tag{2.18}$$

It follows almost immediately that

$$C(x) = A(x)B(x), \tag{2.19}$$

since multiplying the two series $A(x)$ and $B(x)$, and using (2.16), gives the coefficient of x^r as c_r.

In the special case of probability distributions, the probability-generating function of the sum S of two independent non-negative integral-valued random variables X and Y is simply the product of the latter's probability-generating functions.

More generally, we can consider the convolution of several sequences, and the generating function of the convolution is simply the product of the individual generating functions, i.e. the generating function of $\{a_j\}*\{b_j\}*\{c_j\}* \cdots$ is $A(x) B(x) C(x) \cdots$.

If, in particular, we have the sum of several independent random variables of the previous type, e.g.

$$S_n = X_1 + X_2 + \cdots + X_n, \tag{2.20}$$

where the X_k have a common probability distribution given by p_j, with probability-generating function $P(x)$, then the probability-generating function of S_n is $\{P(x)\}^n$. Further, the distribution of S_n is given by a sequence of probabilities which is the n-fold convolution of $\{p_j\}$ with itself. This is written as

$$\{p_j\}*\{p_j\}* \cdots *\{p_j\} = \{p_j\}^{n*}. \qquad (2.21)$$

$$\longleftarrow n \text{ factors } \longrightarrow$$

2.4 Compound distributions

Let us extend the notion at the end of the previous section to the case where the number of random variables contributing to the sum is itself a random variable. Suppose that

$$S_N = X_1 + X_2 + \cdots + X_N, \qquad (2.22)$$

where

$$\left. \begin{array}{l} P\{X_k = j\} = f_j \\ P\{N = n\} = g_n \\ P\{S_N = l\} = h_l \end{array} \right\}. \qquad (2.23)$$

and

Let the corresponding probability-generating functions be

$$\left. \begin{array}{l} F(x) = \sum f_j x^j \\ G(x) = \sum g_n x^n \\ H(x) = \sum h_l x^l \end{array} \right\}. \qquad (2.24)$$

and

Now simple probability considerations show that we can write the probability distribution of S_N as

$$h_l = P\{S_N = l\}$$

$$= \sum_{n=0} P\{N = n\}P\{S_n = l|N = n\}$$

$$= \sum_{n=0} g_n P\{S_n = l|N = n\}. \qquad (2.25)$$

For fixed n, the distribution of S_n is the n-fold convolution of $\{f_j\}$ with itself, i.e. $\{f_j\}^{n*}$. Therefore

$$\sum_{l=0} P\{S_n = l|N = n\}x^l = \{F(x)\}^n. \qquad (2.26)$$

Thus the probability-generating function $H(x)$ can be expressed as

$$H(x) = \sum_{l=0} h_l x^l$$

$$= \sum_{l=0} x^l \sum_{n=0} g_n P\{S_n = l | N = n\}, \quad \text{using (2.25)}$$

$$= \sum_{n=0} g_n \sum_{l=0} P\{S_n = l | N = n\} x^l$$

$$= \sum_{n=0} g_n \{F(x)\}^n, \qquad\qquad \text{using (2.26)}$$

$$= G(F(x)). \tag{2.27}$$

This gives a functionally simple form for the probability-generating function of the *compound distribution* $\{h_l\}$ of the sum S_N.

The formula in (2.27) can be used to obtain several standard results in the theory of distributions. However, in the context of the present book we shall be concerned with applications to a special kind of stochastic process, namely the *branching process* examined in Chapter 6.

2.5 Partial fraction expansions

In many problems some desired probability distribution is obtained by first deriving the corresponding probability-generating function, and then picking out the coefficients in the latter's Taylor expansion. Thus if we have found the probability-generating function $P(x)$, where $P(x) = \sum_{j=0} p_j x^j$, as usual, the coefficients p_j are given by repeated differentiation of $P(x)$, i.e.

$$p_j = \frac{1}{j!} \left(\frac{d}{dx}\right)^j P(x) \Big|_{x=0}$$

$$= P^{(j)}(0)/j!. \tag{2.28}$$

In practice explicit calculation may be very difficult or impossible, and some kind of approximation is desirable. One very convenient method is based on a development in terms of partial fractions.

For the time being let us suppose that the function $P(x)$ is a rational function of the form

$$P(x) = U(x)/V(x), \tag{2.29}$$

where $U(x)$ and $V(x)$ are polynomials without common roots. To begin with we can assume that the degree of $U(x)$ is less than that of $V(x)$, the latter being m, say. Let the equation $V(x) = 0$ have m distinct roots, $x_1, x_2, \cdots, x_m$, so that we can take

$$V(x) = (x - x_1)(x - x_2) \cdots (x - x_m). \tag{2.30}$$

It is a standard result that the function $P(x)$ can be expressed in partial fractions as

$$P(x) = \frac{\rho_1}{x_1 - x} + \frac{\rho_2}{x_2 - x} + \cdots + \frac{\rho_m}{x_m - x}, \tag{2.31}$$

where the constant ρ_1, for example, is given by

$$\rho_1 = \lim_{x \to x_1} (x_1 - x)P(x)$$

$$= \lim_{x \to x_1} \frac{(x_1 - x)U(x)}{V(x)}$$

$$= \frac{-U(x_1)}{(x_1 - x_2)(x_1 - x_3) \cdots (x_1 - x_m)}$$

$$= \frac{-U(x_1)}{V'(x_1)}, \tag{2.32}$$

and in general

$$\rho_j = -U(x_j)/V'(x_j). \tag{2.33}$$

The chief difficulty so far is in obtaining the partial fraction expansion (2.31), and in practice considerable numerical calculation may be required. When, however, the expansion is available, the coefficient p_k of x_k is easily derived. For

$$\frac{1}{x_j - x} = \frac{1}{x_j}\left(1 - \frac{x}{x_j}\right)^{-1}$$

$$= x_j^{-1}\left\{1 + \left(\frac{x}{x_j}\right) + \left(\frac{x}{x_j}\right)^2 + \cdots\right\},$$

and if this expression is substituted in equation (2.31) for $j = 1, 2, \cdots, m$, we can pick out the required coefficient as

$$p_k = \frac{\rho_1}{x_1^{k+1}} + \frac{\rho_2}{x_2^{k+1}} + \cdots + \frac{\rho_m}{x_m^{k+1}}. \tag{2.34}$$

The formula in (2.34) is of course exact, but involves the disadvantage that all m roots of $V(x) = 0$ must be calculated, and this may be prohibitive in practice. Frequently, however, the main contribution comes from a single term. Thus suppose that x_1 is *smaller* in absolute value than all other roots. The first term $\rho_1 x_1^{-k-1}$ will then dominate as k increases. More exactly, we can write

$$p_k \sim \rho_1 x_1^{-k-1} \quad \text{as} \quad k \to \infty. \tag{2.35}$$

This formula is often surprisingly good, even for relatively small values of k. Moreover, if we apply this method to the generating function $Q(x)$, we shall obtain an approximation to the whole tail of the probability distribution beyond some specified point (see Feller, 1957, Chapter 11, for a numerical example).

Most of the restrictions in the foregoing derivation can be relaxed. Thus if $U(x)$ is of degree $m + r$, division by $V(x)$ leads to the form already considered plus a polynomial of degree r, affecting only the first $r + 1$ terms of the distribution. Again if x_h is a double root, then (2.31) contains an additional term of the form $\sigma_h(x - x_h)^{-2}$, which adds a quantity $\sigma_h(k + 1)x_h^{-k-2}$ to the exact expression for p_k in (2.34). Similarly for roots of greater multiplicity. Thus we can see that *provided* x_1 is a *simple* root, smaller in absolute value than all other roots, the asymptotic formula (2.35) will still hold.

It can further be shown that if x_1 is a root of $V(x)$, smaller in absolute value than all other roots, but of multiplicity r, then

$$p_k \sim \binom{k + r - 1}{r - 1}\rho_1 x_1^{-k-r}, \tag{2.36}$$

where now

$$\rho_1 = (-1)^r r!\, U(x_1)/V^{(r)}(x_1). \tag{2.37}$$

As mentioned in Section 2.2 it is sometimes convenient in advanced applications to define the generating function in terms of a complex dummy variable z instead of the real quantity x used above. We can still use the foregoing method of partial fraction expansion for rational functions, but a much wider class of functions in complex variable theory admits of such expansions. Thus suppose that the function $f(z)$ is regular except for poles in any finite region of the z-plane. Then if ζ is not a pole, we have, subject to certain important conditions (see, for example, Copson, 1948, Section 6.8), the result that $f(\zeta)$ is equal to the sum of the residues of $f(z)/(\zeta - z)$ at the poles of $f(z)$. That this does provide a partial fraction expansion is seen from the fact that if α is a pole of $f(z)$ with principle part $\sum_{r=1}^{m} b_r(z - \alpha)^{-r}$, the contribution of α to the required sum is $\sum_{r=1}^{m} b_r(\zeta - \alpha)^{-r}$.

2.6 Moment- and cumulant-generating functions

So far we have been discussing the properties of generating functions, mainly, though not entirely, in the context of probability-generating functions for discrete-valued random variables. There are of course a number of other functions that generate moments and cumulants, and which apply to random variables that are continuous as well as to those that are discrete. Thus we may define the moment-generating function $M(\theta)$ of a random variable X quite generally as

$$M(\theta) = E(e^{\theta X}). \tag{2.38}$$

If X is discrete and takes the value j with probability p_j we have

$$M(\theta) = \sum_j e^{\theta j} p_j \equiv P(e^{\theta}), \tag{2.39}$$

while if X is continuous with a frequency function $f(u)$ we can write

$$M(\theta) = \int_{-\infty}^{\infty} e^{\theta u} f(u) \, du. \tag{2.40}$$

In either case the Taylor expansion of $M(\theta)$ generates the moments given by

$$M(\theta) = 1 + \sum_{r=1}^{\infty} \mu_r' \theta^r / r!, \tag{2.41}$$

where μ_r' is the rth moment about the origin.

Unfortunately, moment-generating functions do not always exist, and it is often better to work with the characteristic function defined by

$$\phi(\theta) = E(e^{i\theta X}), \tag{2.42}$$

which does always exist. The expansion corresponding to (2.41) is then

$$\phi(\theta) = 1 + \sum_{r=1}^{\infty} \mu_r' (i\theta)^r / r!. \tag{2.43}$$

When there is a continuous-frequency function $f(u)$, we have, as the analog of (2.40),

$$\phi(\theta) = \int_{-\infty}^{\infty} e^{i\theta u} f(u) \, du, \tag{2.44}$$

for which there is a simple inversion formula given by the Fourier transform

$$f(u) = \frac{1}{2\pi} \int_{-\infty}^{\infty} e^{-i\theta u} \phi(\theta) \, d\theta. \tag{2.45}$$

It should be noted that this is a complex integral, and usually requires contour integration and the theory of residues for its evaluation. For a fuller account of the theory of characteristic functions, and the appropriate form taken by the inversion theorem under less restrictive conditions than those required above, the reader should consult Parzen (1960, Chapter 9).

For many reasons it often turns out to be simpler to work in terms of cumulants, rather than moments.

The cumulant-generating function is simply the natural logarithm of the moment-generating function or characteristic function, whichever is most convenient. In the former case we have the cumulant-generating function $K(\theta)$ given by

$$K(\theta) = \log M(\theta) \tag{2.46}$$

$$\equiv \sum_{r=1}^{\infty} \kappa_r \theta^r / r!, \tag{2.47}$$

where κ_r is the rth cumulant.

Finally, we note the factorial moment-generating function, sometimes useful in handling discrete variables, defined by

$$Q(y) = P(1 + y) = E\{(1 + y)^j\} \tag{2.48}$$

$$\equiv 1 + \sum_{r=1}^{\infty} \mu'_{[r]} y^r / r!, \tag{2.49}$$

where $\mu'_{[r]}$, the rth factorial moment about the origin, is defined as in (2.13).

PROBLEMS FOR SOLUTION

1. Show that the probability-generating function for the binomial distribution $p_j = \binom{n}{j} p^j q^{n-j}$, where $q = 1 - p$, is $P(x) = (q + px)^n$. Hence prove that the mean is $m = np$, and the variance is $\sigma^2 = npq$.

2. What is the probability-generating function for the Poisson distribution $p_j = e^{-\lambda} \lambda^j / j!, j = 0, 1, 2, \cdots$? Hence show that the mean and variance of the distribution are both equal to λ.

3. Find the probability-generating function of the geometric distribution $p_j = pq^j, j = 0, 1, 2, \cdots$. What are the mean and variance?

4. If $M_X(\theta) = E(e^{\theta X})$ is the moment-generating function of a random variable X, show that the moment-generating function of $Y = (X - a)/b$ is $M_Y(\theta) = e^{-a\theta/b} M_X(\theta/b)$.

5. Using the result of Problem 4, show that the moment-generating function of the binomial variable in Problem 1, measured from its mean value, is $(qe^{-p\theta} + pe^{-q\theta})^n$. Hence calculate the first four moments about the mean.

6. Obtain the probability-generating function of the negative-binomial distribution given by $p_j = \binom{n+j-1}{j} p^n q^j, j = 0, 1, 2, \cdots$. Find the mean and variance.

7. Show that the negative-binomial distribution in Problem 6 is the convolution of n independent geometric distributions, each of the form shown in Problem 3.

8. Show that the cumulant-generating function of the normal distribution given by

$$f(u) = \frac{1}{\sigma(2\pi)^{\frac{1}{2}}} e^{-(u-m)^2/2\sigma^2}, \quad -\infty < u < \infty,$$

is $K(\theta) = m\theta + \tfrac{1}{2}\sigma^2\theta^2.$

9. Suppose that $S_N = X_1 + X_2 + \cdots + X_N$, where the X_j are independent random variables with identical distributions given by

$$P\{X_j = 0\} = q, \; P\{X_j = 1\} = p,$$

and N is a random variable with a Poisson distribution having mean λ. Prove that S_N has a Poisson distribution with mean λp.

CHAPTER 3

Recurrent Events

3.1 Introduction

Some of the simplest kinds of stochastic process occur in relation to a repetitive series of experimental trials such as occur when a coin is tossed over and over again. Thus we might suppose that the chance of heads occurring at any particular trial is p, independently of all previous tosses. It is then not difficult to use standard probability theory to calculate the chance that the accumulated number of heads in n tosses lies between specified limits, or the probability of occurrence of runs of heads of a given length, and so on. We might hesitate to dignify such simple schemes with the title of stochastic process. Nevertheless, with only a little generalization, they provide the basis for a general theory of *recurrent events*. And the importance of this topic in the context of this book is that it provides a number of results and theorems that will be of immediate application to our discussion of Markov chains in Chapter 5. The present chapter introduces the chief concepts required later, and outlines the main results in a comparatively non-rigorous fashion. For a more precise set of definitions and closely reasoned discussion see Chapter 13 of Feller (1957).

3.2 Definitions

Let us suppose that we have a succession of trials, which are *not necessarily independent*, and each of which has a number of possible outcomes, say E_j ($j = 1, 2, \cdots$). We now concentrate attention on some specific event E, which could be one of the outcomes or some more general pattern of outcomes. More generally, we need only consider an event E which may or may not occur at each of a succession of instants, $n = 1, 2, \cdots$.

16

Thus in a sequence of coin-tossing trials E could simply be the occurrence of heads at any particular trial, or it could be the contingency that the accumulated numbers of heads and tails were equal. It is convenient to introduce a mild restriction as to the kind of events we are prepared to consider: namely, that whenever E occurs, we assume the series of trials or observations to start again from scratch for the purposes of looking for another occurrence of E. This ensures that the event "E occurs at the nth trial" depends only on the outcomes of the first n trials, and not on the future, which would lead to unnecessary complications. The practical consequence of this assumption is that if we consider the event E constituted by two successive heads, say HH, then the sequence $HHHT$ (where T stands for tails) shows E appearing only at the second trial. After that, we start again, and the remaining trials HT do not yield E. Similarly, the sequence $HHH|HT|HHH|HHH$ provides just three occurrences of "a run of three heads", as indicated by the divisions shown.

With these definitions, let us specify the probability that E occurs at the nth trial (not necessarily for the first time) as u_n; and the probability that E does occur for the first time at the nth trial as f_n. For convenience we define

$$u_0 \equiv 1, \qquad f_0 \equiv 0, \tag{3.1}$$

and introduce the generating functions

$$U(x) = \sum_{n=0}^{\infty} u_n x^n, \qquad F(x) = \sum_{n=0}^{\infty} f_n x^n. \tag{3.2}$$

Note that the sequence $\{u_n\}$ is *not* a probability distribution, as the probabilities refer to events that are not mutually exclusive. And in many important cases we find $\sum u_n = \infty$. However, the events "E occurs for the first time at the nth trial" are mutually exclusive, and so

$$f = \sum_{n=1}^{\infty} f_n \equiv F(1) \leqslant 1. \tag{3.3}$$

The quantity $1 - f$ can clearly be interpreted as the probability that E does not occur at all in an indefinitely prolonged series of trials.

If $f = 1$, the f_n constitute a genuine set of probabilities, and we can legitimately talk about the mean of this distribution, $\mu = \sum_{n=1}^{\infty} nf_n$. Indeed, even when $f < 1$, we can regard the f_n as providing a probability distribution if we formally assign the value ∞ to the chance of non-occurrence, $1 - f$. In this case we automatically have $\mu = \infty$. The probability distribution f_n therefore refers to the *waiting time* for E, defined as the number of trials up to and including the first occurrence of E; or, more generally, to the *recurrence time* between successive occurrences of E.

Definitions

We call the recurrent event *E persistent* if $f = 1$, and *transient* if $f < 1$.

(There is some variation in the literature regarding the terms employed. The present usage, recommended in Feller (1957, Chapter 13), seems preferable where applications to Markov chains are concerned.)

We call the recurrent event *E periodic* with period t, if there exists an integer $t > 1$ such that E can occur only at trials numbered $t, 2t, 3t, \cdots$ (i.e. $u_n = 0$ whenever n is not divisible by t, and t is the largest integer with this property).

3.3   Basic theorems

The first thing we require is the fundamental relationship between the two sequences $\{u_n\}$ and $\{f_n\}$. This is most easily expressed in terms of the corresponding generating functions. We use the following argument.

To begin with we decompose the event "E occurs at the nth trial", for which the probability is u_n, into the set of mutually exclusive and exhaustive events given by "E occurs for the first time at the jth trial, and again at the nth trial" for $j = 1, 2, 3, \cdots, n$. The probability of the latter is $f_j u_{n-j}$, since the chance that E occurs for the first time at the jth trial is f_j, and, having thus occurred, the chance that it occurs again $n - j$ trials later is u_{n-j}, the two chances being by definition independent. It immediately follows that

$$u_n = f_1 u_{n-1} + f_2 u_{n-2} + \cdots + f_n u_0, \quad n \geqslant 1. \tag{3.4}$$

Now, remembering that $f_0 \equiv 0$, the sequence on the right of (3.4) is, for $n \geqslant 1$, just the convolution $\{u_n\} * \{f_n\}$, for which the generating function is $U(x)F(x)$. But the sequence on the left of (3.4) is, for $n \geqslant 1$, the set of quantities $\{u_n\}$ with the first term $u_0 = 1$ missing. The generating function for this sequence is thus $U(x) - 1$. It follows that $U(x) - 1 = U(x)F(x)$, or

$$U(x) = \frac{1}{1 - F(x)}. \tag{3.5}$$

Suppose now that we regard u_j as the expectation of a random variable that takes the values 1 and 0 as E does or does not occur at the jth trial. The quantity $\sum_{j=1}^{n} u_j$ is thus the expected number of occurrences in n trials. If we define

$$u = \sum_{j=0}^{\infty} u_j, \tag{3.6}$$

then $u - 1$ is the expected number of occurrences in an infinite sequence of trials (using the fact that $u_0 \equiv 1$).

Theorem 3.1

A necessary and sufficient condition for the event E to be persistent is that $\sum u_j$ should diverge.

This theorem is useful for determining the status of a particular kind of event, given the relevant probabilities. The proof is as follows. If E is persistent, then $f = 1$ and $F(x) \to 1$ as $x \to 1$. Thus, from (3.5), $U(x) \to \infty$ and $\sum u_j$ diverges. The converse argument also holds.

Again, if E is transient, $F(x) \to f$ as $x \to 1$. In this case it follows from (3.5) that $U(x) \to (1 - f)^{-1}$. Now $U(x)$ is a series with non-negative coefficients. So by Abel's convergence theorem, the series converges for the value $x = 1$, and has the value $U(1) = (1 - f)^{-1}$. Thus in the transient case we have

$$u = \frac{1}{1-f}, \qquad f = \frac{u-1}{u}. \tag{3.7}$$

A theorem of considerable importance in the theory of recurrent events, and especially in later applications to Markov chains, is the following.

Theorem 3.2

Let E be a persistent, not periodic, recurrent event with mean recurrence time $\mu = \sum nf_n = F'(1)$, then

$$u_n \to \mu^{-1} \quad \text{as} \quad n \to \infty. \tag{3.8}$$

In particular $u_n \to 0$ if $\mu = \infty$.

We shall not give the proof here: it can be carried through in a relatively elementary way, but is rather lengthy (see, for example, Feller, 1957, Chapter 13, Section 10). The corresponding result for periodic events (which can be deduced quite easily from Theorem 3.2) is:

Theorem 3.3

If E is persistent with period t, then

$$u_{nt} \to t\mu^{-1} \quad \text{as} \quad n \to \infty, \tag{3.9}$$

and $u_k = 0$ for every k not divisible by t.

3.4 Illustration

As an illustration of some of the foregoing ideas let us consider a series of independent trials, at each of which there is a result that is either a success or failure, the probabilities being p and $q = 1 - p$, respectively. Let E be the event "the cumulative numbers of successes and failures are equal". We might suppose for example that a gambler wins or loses a unit sum of money at each trial with chances p or q, and E represents the contingency that his accumulated losses and gains are exactly zero.

Clearly, the numbers of successes and failures can only be equal at even-numbered trials, i.e.

$$u_{2n} = \binom{2n}{n} p^n q^n, \qquad u_{2n+1} = 0. \qquad (3.10)$$

Using Stirling's approximation to large factorials, we can write

$$u_{2n} \sim \frac{(4pq)^n}{(n\pi)^{\frac{1}{2}}}. \qquad (3.11)$$

Thus if $p \neq \frac{1}{2}$, we have $4pq < 1$ and $\sum u_{2n}$ converges faster than the geometric series with ratio $4pq$. According to Theorem 3.1 the event E cannot be persistent, and is therefore transient. If, on the other hand, $p = q = \frac{1}{2}$, we have $u_{2n} \sim (n\pi)^{-\frac{1}{2}}$ and $\sum u_{2n}$ diverges, although $u_{2n} \to 0$ as $n \to \infty$. In this case E is persistent, but the mean recurrence time will be infinite (the latter follows directly from Theorem 3.2, but a direct proof is given below).

In terms of the gambling interpretation, we could say that when $p > \frac{1}{2}$ there is probability one that the accumulated gains and losses will be zero only a finite number of times: the game is favorable, and after some initial fluctuation the net gain will remain positive. But if $p = \frac{1}{2}$, the situation of net zero gain will recur infinitely often.

An exact calculation of the various properties of the process and of the distribution of recurrence times is easily effected using the generating functions of the previous section. First, from (3.10) we have

$$U(x) = \sum_{n=0}^{\infty} u_{2n} x^{2n}$$

$$= \sum_{n=0}^{\infty} \binom{2n}{n} (pqx^2)^n$$

$$= (1 - 4pqx^2)^{-\frac{1}{2}}. \qquad (3.12)$$

If $p \neq \frac{1}{2}$, we have

$$u = U(1) = (1 - 4pq)^{-\frac{1}{2}} = |p - q|^{-1}$$

so that the expected number of occurrences of E, i.e. of returns to net zero gain is

$$u - 1 = \frac{1}{|p - q|} - 1, \qquad (3.13)$$

and the chance that E will occur at least once is

$$f = \frac{u-1}{u} = 1 - |p - q|.$$ (3.14)

We also have

$$F(x) = \frac{U(x) - 1}{U(x)} = 1 - (1 - 4pqx^2)^{\frac{1}{2}}.$$ (3.15)

When $p = \frac{1}{2}$, this becomes

$$F(x) = 1 - (1 - x^2)^{\frac{1}{2}},$$ (3.16)

from which we obtain the probability distribution.

$$\left. \begin{array}{l} f_{2n} = \binom{2n-2}{n-1} \dfrac{1}{n2^{2n-1}}, \quad n \geqslant 1 \\[2mm] f_{2n+1} = 0, \qquad\qquad\qquad\quad n \geqslant 0 \end{array} \right\}.$$ (3.17)

The mean value of this distribution is of course $\mu = F'(1) = \infty$, as already remarked above.

3.5 Delayed recurrent events

It is convenient to introduce at this point a certain extension of the above account of recurrent events, mainly because we shall want to make use of the appropriate modification of Theorem 3.2 in our discussion of Markov chains in Chapter 5. Actually, the whole subject of recurrent events can be treated as a special case of the more general theory of *renewal processes*. For an extensive account of this topic see the recent book by D. R. Cox (1962).

In formulating the basic model for ordinary recurrent events in Section 3.2, we used the sequence $\{f_n\}$ to represent the probability distribution of recurrence times between successive occurrences of the event E, *including* the time up to the first appearance of E. We now modify this model by supposing that the probability distribution of the time up to the first occurrence of E is given by the sequence $\{b_n\}$, which is in general different from $\{f_n\}$. It is as though we missed the beginning of the whole series of trials, and only started recording at some later point. Following Feller (1957, Chapter 13, Section 5), we shall call E a *delayed recurrent event*. The basic equation for the corresponding process is obtained by the following modification of the earlier argument.

As before, we decompose the event "E occurs at the nth trial", for which the probability is u_n, into the set of mutually exclusive and exhaustive events given by "E occurs for the first time at the jth trial, and again at the nth trial" for $j = 1, 2, 3, \cdots, n$. The probability of the latter is $f_j u_{n-j}$ for $1 \leqslant j \leqslant n - 1$, but b_n for $j = n$. Hence this time we have

$$u_n = f_1 u_{n-1} + f_2 u_{n-2} + \cdots + f_{n-1} u_1 + b_n, \quad n \geqslant 1. \tag{3.18}$$

For delayed events it is convenient to adopt the convention that

$$u_0 = f_0 = b_0 = 0. \tag{3.19}$$

Equation (3.18) can then be written in terms of the sequences $\{u_n\}$, $\{f_n\}$ and $\{b_n\}$ as

$$\{u_n\} = \{u_n\} * \{f_n\} + \{b_n\}. \tag{3.20}$$

If the corresponding generating functions are

$$U(x) = \sum_{n=0}^{\infty} u_n x^n, \qquad F(x) = \sum_{n=0}^{\infty} f_n x^n, \qquad B(x) = \sum_{n=0}^{\infty} b_n x^n, \tag{3.21}$$

equation (3.20) leads immediately to

$$U(x) = U(x)F(x) + B(x),$$

or

$$U(x) = \frac{B(x)}{1 - F(x)}, \tag{3.22}$$

which is the required extension of (3.5). (Alternatively, we could simply multiply (3.18) by x^n, and sum for $n = 0, 1, 2, \cdots$.)

The theorem we require is the following appropriate extension of Theorem 3.2, which we simply quote here without proof (see Feller, 1957, Chapter 13, Sections 4 and 5).

Theorem 3.4

Let E be a persistent and not periodic delayed recurrent event, with mean recurrence time $\mu = \sum n f_n = F'(1)$, then

$$u_n \to \mu^{-1} \sum b_n \quad \text{as} \quad n \to \infty. \tag{3.23}$$

In the case when E is transient, and $f = \sum f_n < 1$, we can easily show that

$$u = \sum u_n = (1 - f)^{-1} \sum b_n, \tag{3.24}$$

which is the obvious extension of the result in equation (3.7).

PROBLEMS FOR SOLUTION

1. Consider a series of independent trials, at each of which there is success or failure, with probabilities p and $q = 1 - p$, respectively. Let E be the very simple

event "success". Find the generating functions $U(x)$ and $F(x)$. Hence show that the number of trials between consecutive successes has a geometric distribution.

2. Suppose that two distinguishable unbiased coins are tossed repeatedly. Let E be the event "the accumulated number of heads is the same for each coin". Prove that E is persistent with an infinite mean recurrence time.

3. An unbiased die is thrown successively. The event E is defined by "all six spot-numbers have appeared in equal numbers". Show that E is periodic and transient.

4. Suppose we have a sequence of mutually independent random variables X_k with a common probability distribution given by

$$P\{X_k = a\} = \frac{b}{a+b}, \qquad P\{X_k = -b\} = \frac{a}{a+b},$$

where a and b are positive integers. Define the sum S_n by

$$S_n = X_1 + X_2 + \cdots + X_n.$$

Show that the event "$S_n = 0$" is persistent.

5. Suppose a gambler wins or loses a unit sum at each of a series of independent trials with equal probability. Let E be the event "the gambler's net gain is zero after the present trial, but was negative after the last trial". What is the distribution of the recurrence time of this event?

CHAPTER 4

Random Walk Models

4.1 Introduction

There are often several approaches that can be tried for the investigation of a stochastic process. One of these is to formulate the basic model in terms of the motion of a particle which moves in discrete jumps with certain probabilities from point to point. At its simplest, we imagine the particle starting at the point $x = k$ on the x-axis at time $t = 0$, and at each subsequent time $t = 1, 2, \cdots$ it moves one unit to the right or left with probabilities p or $q = 1 - p$, respectively. A model of this kind is used in physics as a crude approximation to one-dimensional diffusion or Brownian motion, where a particle may be supposed to be subject to a large number of random molecular collisions. With equal probabilities of a transition to left or right, i.e. $p = q = \frac{1}{2}$, we say the walk is *symmetric*. If $p > \frac{1}{2}$, there is a *drift* to the right; and if $p < \frac{1}{2}$ the drift is to the left.

Such a model can clearly be used to represent the progress of the fortunes of the gambler in Section 3.4, who at each trial won or lost one unit with chance p or q, respectively. The actual position of the particle on the x-axis at any time indicates the gambler's accumulated loss or gain.

In the diffusion application we might wish to have no limits on the extent to which the particle might move from the origin. We should then say that the walk was *unrestricted*. If, however, we wanted the position of the particle to represent the size of a population of individuals, where a step to the right corresponded to a birth and a step to the left meant a death, then the process would have to stop if the particle ever reached the origin $x = 0$. The random walk would then be *restricted*, and we should say that $x = 0$ was an *absorbing barrier*.

Again, to return to the gambler, suppose that the latter starts with a capital of amount k and plays against an adversary who has an amount $a - k$, so that the combined capital is a. Then the game will cease if, either $x = 0$ and the gambler has lost all his money, or $x = a$ and the gambler has won all his opponent's money. In this case there are absorbing barriers at $x = 0$ and $x = a$.

As an extension of these ideas we can also take into account the possibility of the particle's remaining where it is. We could also introduce transitions from any point on the line to any other point, and not merely to those in positions one unit to the left or right. Other modifications are the adoption of *reflecting barriers*. Thus if $x = 0$ is a reflecting barrier, a particle at $x = 1$ moves to $x = 2$ with probability p, and remains at $x = 1$ with probability q (i.e. it moves to $x = 0$ and is immediately reflected back to $x = 1$). Further, we can have an *elastic barrier*, in which the reflecting barrier in the scheme just mentioned is modified so as either to reflect or to absorb with probabilities δq and $(1 - \delta)q$, respectively.

4.2 Gambler's ruin

Historically, the classical random walk has been the *gambler's ruin problem*. An analysis of the main features of this problem will illustrate a number of techniques that may be of use in the investigation of more difficult situations.

Let us suppose, as suggested in the previous section, that we have a gambler with initial capital k. He plays against an opponent whose initial capital is $a - k$. The game proceeds by stages, and at each step the first gambler has a chance p of winning one unit from his adversary, and a chance $q = 1 - p$ of losing one unit to his adversary. The actual capital possessed by the first gambler is thus represented by a random walk on the non-negative integers with absorbing barriers at $x = 0$ and $x = a$, absorption being interpreted as the ruin of one or the other of the gamblers.

Probability of ruin

Now, the probability q_k of the first gambler's ruin when he starts with an initial capital of k can be obtained by quite elementary considerations as follows. After the first trial this gambler's capital is either $k + 1$ or $k - 1$, according as he wins or loses that game. Hence

$$q_k = pq_{k+1} + qq_{k-1}, \quad 1 < k < a - 1,$$
$$q_1 = pq_2 + q, \quad k = 1,$$
$$q_{a-1} = qq_{a-2}, \quad k = a - 1,$$

all of which equations can be subsumed under the general form

$$q_k = pq_{k+1} + qq_{k-1}, \quad 1 \leqslant k \leqslant a - 1, \tag{4.1}$$

provided we adopt the conventions that

$$q_0 = 1, \qquad q_a = 0. \tag{4.2}$$

The difference equation in (4.1) is easily solved by a standard method, e.g. Goldberg (1958, Section 3.3). If we put $q_k = w^k$ in (4.1) we obtain the auxiliary equation

$$pw^2 - w + q = 0, \tag{4.3}$$

which has roots $w = 1, q/p$.

Thus if $p \neq \frac{1}{2}$ the equation (4.3) has two different roots, and we can write the general solution of (4.1) as

$$q_k = A(1)^k + B(q/p)^k, \tag{4.4}$$

where A and B are constants which must be chosen to satisfy the boundary conditions appearing in (4.2). We therefore have the two equations

$$1 = A + B, \qquad 0 = A + B(q/p)^a,$$

with solutions

$$A = \frac{(q/p)^a}{(q/p)^a - 1}, \qquad B = -\frac{1}{(q/p)^a - 1}. \tag{4.5}$$

The required solution for q_k is thus

$$q_k = \frac{(q/p)^a - (q/p)^k}{(q/p)^a - 1}, \tag{4.6}$$

and it is not difficult to show that this is unique.

In a similar way we can calculate the probability p_k of the first gambler's success, i.e. his opponent's ruin. We merely have to interchange p and q, and write $a - k$ for k. These substitutions applied to formula (4.6) then yield, after rearrangement,

$$p_k = \frac{(q/p)^k - 1}{(q/p)^a - 1}. \tag{4.7}$$

It now follows from (4.6) and (4.7) that

$$p_k + q_k = 1, \tag{4.8}$$

so that the chance of an unending contest between the two gamblers is zero, a fact we were careful not to assume in advance.

If $p = \frac{1}{2}$, the auxiliary equation has two equal roots $w = 1$, and the general solution must be written in the form

$$q_k = C(1)^k + Dk(1)^k. \tag{4.9}$$

Using the boundary conditions to determine the constants C and D leads to the final result

$$q_k = 1 - k/a. \tag{4.10}$$

(Alternatively, we could let $p \to \frac{1}{2}$ in (4.6) and use l'Hôpital's rule on the right-hand side.) Similarly, we find

$$p_k = k/a, \tag{4.11}$$

and so (4.8) holds as before.

It is worth remarking at this point that the difference equation in (4.1) can also be solved by a generating-function method in which we put $Q(x) = \sum_{k=0}^{a} q_k x^k$, but the above derivation is somewhat shorter.

The implications of these results for gamblers are discussed at length in Feller (1957, Chapter 14). It follows, for example, from (4.11) that given equal skill (i.e. $p = \frac{1}{2}$) a gambler's chance of winning all his adversary's money is proportional to his share of their total wealth. However, we are more concerned here with the methods of analysis of the random walk model.

Expected duration of game

The expected duration of the game can also be obtained in a comparatively elementary way, without recourse to a full discussion of the whole probability distribution, undertaken in the next section. It is worth emphasizing the importance of shortcuts of this kind, which are frequently possible if we are content to evaluate mean values only.

Suppose the duration of the game has a finite expectation d_k, starting the random walk as before from the point $x = k$. If the first trial leads to a win for the first gambler, the conditional duration from that point on is d_{k+1}. The expected duration of the whole game if the first trial is a win (occurring with probability p) is thus $1 + d_{k+1}$. Similarly, the expected duration of the whole game if the first trial is a loss (occurring with probability q) is $1 + d_{k-1}$. Therefore

$$d_k = p(1 + d_{k+1}) + q(1 + d_{k-1})$$

$$= 1 + pd_{k+1} + qd_{k-1}, \qquad 1 \leqslant k \leqslant a - 1, \tag{4.12}$$

the extreme values $k = 1$ and $k = a - 1$ being included on the assumptions

$$d_0 = 0, \qquad d_a = 0. \tag{4.13}$$

Now the difference equation (4.12) is simply a non-homogeneous version of (4.1) with d_k written for q_k. We can, therefore, utilize the previous general solution (4.4) provided we add on a particular solution of (4.12). Inspection of (4.12) suggests that a solution of the form $d_k = \lambda k$ might be possible. Actual substitution shows that $\lambda = (q - p)^{-1}$. The general solution of (4.12) is, accordingly, for $p \neq \frac{1}{2}$,

$$d_k = \frac{k}{q - p} + A + B\left(\frac{q}{p}\right)^k.$$ (4.14)

The constants A and B are easily evaluated using the boundary conditions (4.13). The final form is then

$$d_k = \frac{k}{q - p} - \frac{a}{q - p}\frac{1 - (q/p)^k}{1 - (q/p)^a}.$$ (4.15)

If $p = \frac{1}{2}$, we can either let $p \to \frac{1}{2}$ in (4.15), or solve from the beginning, using the general solution of the homogeneous equation, i.e. the result in (4.9). This time the particular solution $d_k = k/(q - p)$ breaks down, but we easily obtain another particular solution $d_k = -k^2$. In either case, we obtain the result for $p = \frac{1}{2}$ as

$$d_k = k(a - k).$$ (4.16)

It is interesting to observe that these results imply that expected durations are much longer than one might intuitively suppose. Thus suppose that the two gamblers are of equal skill, i.e. $p = \frac{1}{2}$, and that they start with equal amounts of capital, $k = 500$ and $a = 1000$, say. Then it can be seen from (4.16) that the average duration of the game is a quarter of a million trials! Even with $k = 1$, and $a = 1000$, we have $d_k = 999$.

4.3 Probability distribution of ruin at nth trial

In the previous section we derived two special implications of the gambler's ruin model, namely the probability of ruin and the expected duration of the game, by relatively simple means. Let us now determine the whole probability distribution for the termination of the process at any particular trial. We shall write q_{kn} for the probability that the process ends with the nth trial at $x = 0$ (i.e. the first gambler's ruin at the nth step).

Suppose that ruin actually occurs at the $(n + 1)$th trial, the initial capital being k, for which the appropriate probability is $q_{k,n+1}$. If the first trial is a win, with chance p, the capital is increased to $k + 1$, and the conditional probability of ruin at the $(n + 1)$th step from the beginning is now $q_{k+1,n}$. Similarly if the first trial is a loss, with chance q, the capital is decreased to $k - 1$, and the conditional probability of ruin in n further steps is $q_{k-1,n}$. We can thus form the difference equation

$$q_{k,n+1} = pq_{k+1,n} + qq_{k-1,n},\qquad (4.17)$$

involving the two variables k and n subject to

$$1 < k < a - 1, \qquad n \geqslant 1.$$

We can use (4.17) over the extended range

$$1 \leqslant k \leqslant a - 1, \qquad n \geqslant 0 \qquad (4.18)$$

by writing

$$\left. \begin{array}{l} q_{0n} = q_{an} = 0 \quad \text{for} \quad n \geqslant 1 \\ q_{00} = 1; q_{k0} = 0 \quad \text{for} \quad k > 0 \end{array} \right\}.\qquad (4.19)$$

We now introduce a generating function for variations in n, namely

$$Q_k(x) = \sum_{n=0}^{\infty} q_{kn}x^n.\qquad (4.20)$$

Multiplying (4.17) by x^{n+1} and adding for $n = 0, 1, 2, \cdots$ gives

$$Q_k(x) = pxQ_{k+1}(x) + qxQ_{k-1}(x),\qquad (4.21)$$

valid for k in the range given by (4.18). From the boundary conditions in (4.19) we can also obtain

$$Q_0(x) = 1, \qquad Q_a(x) = 0.\qquad (4.22)$$

Now (4.21) is a homogeneous difference equation for Q_k, of a type similar to that in equation (4.1). For the purposes of solving the present equation, x can be regarded as a constant. This time, however, the coefficients on the right-hand side of the equation do not sum to unity, so we cannot use the previous solution as it stands.

Following the previous method of solution, we put $Q_k = w^k$ in (4.21) to obtain the auxiliary equation

$$pxw^2 - w + qx = 0.\qquad (4.23)$$

The roots of this quadratic in w are

$$w_1 = \frac{1 + (1 - 4pqx^2)^{\frac{1}{2}}}{2px}, \qquad w_2 = \frac{1 - (1 - 4pqx^2)^{\frac{1}{2}}}{2px}.\qquad (4.24)$$

We can now write the required general solution in the form

$$Q_k = Aw_1^{\,k} + Bw_2^{\,k},\qquad (4.25)$$

where the boundary conditions (4.22) will enable us to calculate the quantities A and B, which are of course functions of x. After a little reduction, we find

$$Q_k(x) = \left(\frac{q}{p}\right)^k \frac{w_1^{a-k}(x) - w_2^{a-k}(x)}{w_1^a(x) - w_2^a(x)} \qquad (4.26)$$

(using, for example, the fact that $w_1(x)w_2(x) = q/p$), and this is the generating function for the probability of absorption at $x = 0$. The corresponding generating function $P_k(x)$ for the probability p_{kn} of absorption at $x = a$ at the nth trial is clearly obtained by replacing p, q, and k in (4.26) by q, p, and $a - k$, respectively. Finally, the sum of these two generating functions, $Q_k(x) + P_k(x)$, must give the generating function of the probability that the process terminates at the nth trial (either at $x = 0$ or $x = a$).

Special case $a = \infty$

Let us now consider for a moment the special case $a = \infty$, which corresponds to having an absorbing barrier only at $x = 0$. In the gambling model we can imagine the case of a gambler playing against an infinitely rich adversary. It can be seen from (4.23) and (4.24) that, for $|x| < 1$, the roots w_1 and w_2 satisfy the inequality $0 < w_2 < 1 < w_1$. If, therefore, we let $a \to \infty$ in (4.26), we obtain the solution

$$Q_k(x) = (q/p)^k w_1^{-k}(x)$$
$$= w_2^k(x)$$
$$= \left\{ \frac{1 - (1 - 4pqx^2)^{\frac{1}{2}}}{2px} \right\}^k. \qquad (4.27)$$

Alternatively, we can solve the basic difference equation subject only to the boundary condition $Q_0(x) = 1$. The solution will be unbounded unless $A = 0$. Hence the final result shown in (4.27).

Now $Q_k(x)$ in (4.27) is the generating function for the probability that, starting from the point $x = 0$, the particle will be absorbed at the origin on exactly the nth step. Alternatively, if we choose to envisage a completely unrestricted random walk starting from $x = k$, $Q_k(x)$ is the generating function for the distribution of *first-passage times* through a point k units to left of the initial position. First-passage time distributions for points to the right of the initial point are given simply by interchanging p and q, and the appropriate generating function comes out to be $w_1^{-k}(x)$ instead of $w_2^k(x)$. It will be noticed that $w_1^{-k}(x)$ is the kth power of the generating function for the first-passage time to a point just one unit to the right of the starting point. This is what we should expect, since a little reflection shows that the first-passage time to a point k units to the right will be the sum of k independent first-passage times for successive displacements each of one unit to the right.

If we put $x = 1$ in (4.27) we obtain the total chance of ruin at some stage when playing against an infinitely rich opponent, i.e.

$$q_k = \sum_{n=0}^{\infty} q_{kn} = Q_k(1) = \left\{ \frac{1 - |p - q|}{2p} \right\}^k. \tag{4.28}$$

Thus

$$\begin{aligned} q_k &= (q/p)^k, & p > q \\ &= 1, & p \leqslant q \end{aligned}, \tag{4.29}$$

a result which could have been obtained directly from (4.6) by letting $a \to \infty$.

Exact values of probabilities in general case

Let us now return to the general case with two finite absorbing barriers at $x = 0$ and $x = a$, and see how an explicit expression for q_{kn} may be derived. In principle we merely pick out the coefficient of x^n in the generating function $Q_k(x)$ shown in (4.26), but a little care is needed in actually doing this in practice.

First, we notice that $Q_k(x)$ appears to involve the quantity $(1 - 4pqx^2)^{\frac{1}{2}}$ through the roots w_1 and w_2. It is easily shown, however, that $w_1{}^r - w_2{}^r$ (for integral r) is a rational function in x multiplied by $(1 - 4pqx^2)^{\frac{1}{2}}$. The latter quantity appears in both numerator and denominator, and so cancels out to leave $Q_k(x)$ as the ratio of two polynomials, where the degree of the denominator is $a - 1$ for a odd and $a - 2$ for a even; and the degree of the numerator is $a - 1$ for $a - k$ odd and $a - 2$ for $a - k$ even. Thus the degree of the numerator can never exceed that of the denominator by more than one unit.

For $n > 1$ we could therefore calculate q_{kn} using the kind of partial fraction expansion already discussed in Chapter 2. But the algebra simplifies considerably if we first introduce a new variable θ given by

$$\frac{1}{\cos \theta} = 2(pq)^{\frac{1}{2}}x. \tag{4.30}$$

We can easily find

$$1 \pm (1 - 4pqx^2)^{\frac{1}{2}} = \frac{\cos \theta \pm i \sin \theta}{\cos \theta},$$

and

$$2px = \left(\frac{p}{q}\right)^{\frac{1}{2}} \frac{1}{\cos \theta}.$$

Hence

$$w_{1,2}(x) = (q/p)^{\frac{1}{2}}(\cos \theta \pm i \sin \theta)$$

$$= (q/p)^{\frac{1}{2}} e^{\pm i\theta}. \tag{4.31}$$

We can now write

$$Q_k(x) = \left(\frac{q}{p}\right)^k \frac{(q/p)^{\frac{1}{2}(a-k)}(e^{(a-k)i\theta} - e^{-(a-k)i\theta})}{(q/p)^{\frac{1}{2}a}(e^{ai\theta} - e^{-ai\theta})}$$

$$= \left(\frac{q}{p}\right)^{\frac{1}{2}k} \frac{\sin(a-k)\theta}{\sin a\theta}. \tag{4.32}$$

Now the roots of the denominator in (4.32) are $\theta = 0$, π/a, $2\pi/a$, $\cdots$, and the corresponding values of x are given by

$$x_j = \frac{1}{2(pq)^{\frac{1}{2}} \cos j\pi/a}. \tag{4.33}$$

All possible values of x_j are obtained by putting $j = 0, 1, \cdots, a$. But to $j = 0$ and $j = a$ there correspond the values $\theta = 0$ and $\theta = \pi$, which are also roots of the numerator in (4.32). Further, if a is even, there is no x_j corresponding to $j = \frac{1}{2}a$. Thus if a is odd we have all $a - 1$ roots x_j with $j = 1, 2, \cdots, a - 1$; but if a is even, the value $j = \frac{1}{2}a$ must be omitted.

We can now write

$$Q_k(x) = Ax + B + \frac{\rho_1}{x_1 - x} + \cdots + \frac{\rho_{a-1}}{x_{a-1} - x}, \tag{4.34}$$

where

$$\rho_j = \lim_{x \to x_j} (x_j - x)\left(\frac{q}{p}\right)^{\frac{1}{2}k} \frac{\sin(a-k)\theta}{\sin a\theta}$$

$$= \left(\frac{q}{p}\right)^{\frac{1}{2}k} \frac{\sin\{(a-k)j\pi/a\}}{-a \cos j\pi \cdot (d\theta/dx)_{x=x_j}},$$

using l'Hôpital's rule for the limit of $(x_j - x)/\sin a\theta$,

$$= -\left(\frac{q}{p}\right)^{\frac{1}{2}k} \frac{\sin\{(a-k)j\pi/a\}\sin j\pi/a}{2a(pq)^{\frac{1}{2}} \cos j\pi \cos^2 j\pi/a}$$

$$= \left(\frac{q}{p}\right)^{\frac{1}{2}k} \frac{\sin kj\pi/a \sin j\pi/a}{2a(pq)^{\frac{1}{2}} \cos^2 j\pi/a}. \tag{4.35}$$

Finally, we can pick out the coefficient q_{kn} of x^n in the partial fraction expansion (4.34), for $n > 1$, as

$$q_{kn} = a^{-1}2^n p^{\frac{1}{2}(n-k)} q^{\frac{1}{2}(n+k)} \sum_{j=1}^{a-1} \cos^{n-1} j\pi/a \sin j\pi/a \sin kj\pi/a, \quad (4.36)$$

using the method previously described in Section 2.5.

The special case $a = \infty$ can be dealt with more simply, since we have only to expand the simpler form given by the generating function in (4.27). We can do this most easily by using Lagrange's expansion (see, for example, Copson, 1948, Section 6.23). This expansion is often very useful and easy to handle, but seems not to be as widely known as it should be. It is therefore worth quoting the main theorem required in some detail, though the interested reader should consult the reference given for fuller explanations.

Lagrange's formula for the reversion of series

Suppose $f(w)$ is a function of the complex variable w, regular in the neighborhood of w_0, and $f(w_0) = z_0$, where $f'(w_0) \neq 0$, then the equation

$$f(w) = z \quad (4.37)$$

has a unique solution, regular in a neighborhood of z_0, given by

$$w = w_0 + \sum_{n=1}^{\infty} \frac{(z - z_0)^n}{n!} \left[\frac{d^{n-1}}{dw^{n-1}} \{\phi(w)\}^n \right]_{w=w_0}, \quad (4.38)$$

where
$$f(w) - z_0 = \frac{w - w_0}{\phi(w)}. \quad (4.39)$$

More generally, we can obtain the expansion of a function $\psi(w)$ as

$$\psi(w) = \psi(w_0) + \sum_{n=1}^{\infty} \frac{(z - z_0)^n}{n!} \left[\frac{d^{n-1}}{dw^{n-1}} \psi'(w)\{\phi(w)\}^n \right]_{w=w_0} \quad (4.40)$$

We can apply this theorem to (4.27) quite easily. Treating x as a complex variable, we want to expand the function

$$\psi(w) = Q_k(x) = w^k \quad (4.41)$$

in powers of x, where w is the root of (4.23) that vanishes when $x = 0$, i.e. $w_0 = 0$, $x_0 = 0$. Now we can rewrite (4.23) as

$$x = \frac{w}{pw^2 + q}. \quad (4.42)$$

Hence, in the notation of the above theorem, we have

$$\phi(w) = pw^2 + q. \quad (4.43)$$

The required expansion is therefore obtained immediately from (4.40) as

$$Q_k(x) = w^k = \sum_{n=1}^{\infty} \left[\frac{d^{n-1}}{dw^{n-1}} kw^{k-1}(pw^2 + q)^n \right]_{w=0}. \qquad (4.44)$$

Now if we evaluate the coefficient in square brackets on the right of (4.44) by using the usual formula for the repeated differentiation of a product, we see that the only term surviving when $w = 0$ arises from differentiating w^{k-1} exactly $k - 1$ times; the quantity $(pw^2 + q)^n$ must then be differentiated $n - k$ times, and the only surviving term here comes from the coefficient of $(w^2)^{\frac{1}{2}(n-k)}$. Collecting terms gives the required coefficient of x^n as

$$q_{kn} = \frac{k}{n} \binom{n}{\frac{1}{2}n - \frac{1}{2}k} p^{\frac{1}{2}(n-k)} q^{\frac{1}{2}(n+k)}, \qquad (4.45)$$

where we consider only those terms for which $\frac{1}{2}(n - k)$ is an integer in the interval $[0, n]$.

4.4 Extensions

So far in this chapter, we have restricted the discussion to certain special kinds of one-dimensional random walk models. For example, we have supposed that the only transitions possible at any trial are from the point in question to one of the points just one unit to the left or right. But, more generally, we could consider the possibility of transitions to any other point, such as occur in applications to sequential sampling procedures (see Feller, 1957, Chapter 14, Section 8).

Again we can consider extensions to two or more dimensions, and we shall have occasion later in this book to see how certain aspects of rather complicated situations can sometimes be represented in this way. Consider, for example, the epidemic model of Section 12.3, where the numbers of susceptibles and infectives in the population at any moment may be represented by a point in the plane with coordinates (r, s). Only certain kinds of transitions, with specified probabilities, are possible, and there are two barriers, the one given by $s = 0$ obviously being absorbing (since the epidemic ceases when there are no more infectives in circulation).

We shall not develop any special theory here for such extensions, but shall merely note an interesting result due to Pólya showing an important difference of behavior for an unrestricted symmetric random walk in three dimensions compared with the corresponding results in one and two dimensions.

The case of an unrestricted symmetric random walk in one dimension has already been treated in Section 3.4 in the context of recurrent event theory. We note that for $p = \frac{1}{2}$ there is probability one that the particle will return to the origin sooner or later, i.e. the event is persistent, though it was shown that the mean recurrence time was infinite.

Now consider the case of a symmetric random walk in two dimensions where the particle at the point (x, y) has a probability of $\frac{1}{4}$ of moving at the next step to any one of its four neighbors $(x + 1, y)$, $(x - 1, y)$, $(x, y + 1)$, $(x, y - 1)$. Let u_n be the chance that a particle starting at the origin, is again at the origin at the nth step. This can only happen if the numbers of steps in each of the positive x and y directions equal those in the negative x and y directions, respectively. Thus $u_{2n+1} = 0$, and

$$u_{2n} = \frac{1}{4^{2n}} \sum_{k=0}^{n} \frac{(2n)!}{k!\,k!\,(n-k)!\,(n-k)!}$$

$$= \frac{1}{4^{2n}} \binom{2n}{n} \sum_{k=0}^{n} \binom{n}{k}^2. \tag{4.46}$$

The first expression above follows from the appropriate multinomial distribution in which we have k steps in each of the two x directions, and $n - k$ steps in each of the two y directions. The second expression may be developed, using a well-known identity for binomial coefficients, namely,

$$\sum_{k=0}^{n} \binom{n}{k}^2 = \binom{n}{0}\binom{n}{n} + \binom{n}{1}\binom{n}{n-1} + \cdots$$

$$= \text{coeff. of } x^n \text{ in } \left\{ \binom{n}{0} + \binom{n}{1}x + \cdots + \binom{n}{n}x^n \right\}^2$$

$$= \text{coeff. of } x^n \text{ in } (1 + x)^{2n}$$

$$= \binom{2n}{n}.$$

Therefore

$$u_{2n} = 4^{-2n}\binom{2n}{n}^2$$

$$\sim \frac{1}{\pi n}, \tag{4.47}$$

using Stirling's approximation to the factorials. It is evident from (4.47) that $\sum u_{2n}$ diverges so that a return to the origin is, once more, a persistent event.

A different state of affairs results, however, when we go up to three dimensions. We now imagine a particle moving on a cubic lattice, the chance of moving to any one of the six neighbors of a given point being $\frac{1}{6}$. The expression corresponding to (4.46) above is easily obtained as

$$u_{2n} = \frac{1}{6^{2n}} \sum_{j+k \leqslant n} \frac{(2n)!}{j!\,j!\,k!\,k!\,(n-j-k)!\,(n-j-k)!}$$

$$= \frac{1}{2^{2n}} \binom{2n}{n} \sum_{j+k \leqslant n} \left\{ \frac{1}{3^n} \frac{n!}{j!\,k!\,(n-j-k)!} \right\}^2. \qquad (4.48)$$

Now the term in the braces is part of a trinomial distribution, P_r, say, where $\sum_r P_r = 1$. Thus

$$\sum P_r^2 \leqslant \max\{P_r\} \cdot \sum P_r = \max\{P_r\}.$$

But the maximum value of P_r occurs when j and k are both near to $\frac{1}{3}n$. Hence, again using Stirling's approximation to the factorials, we easily obtain

$$u_{2n} \sim Cn^{-\frac{3}{2}}, \qquad (4.49)$$

where C is a constant independent of n. This time we see that $\sum u_{2n}$ converges, so that in three dimensions a return to the origin is a transient event. The chance of a return has been computed to be approximately 0.35.

PROBLEMS FOR SOLUTION

1. Examine the gambler's ruin problem of Sections 4.1 and 4.2, where the chances of winning and losing one unit are p and q, respectively; and the chance of a tie or draw is r ($p+q+r=1$). Show that the probability of ruin is

$$\frac{(q/p)^a - (q/p)^k}{(q/p)^a - 1},$$

where the gambler's initial capital is k and his opponent has $a-k$; and that the expected duration of the game is

$$\frac{1}{(q-p)(1-r)} \left\{ k - a \frac{(q/p)^k - 1}{(q/p)^a - 1} \right\}.$$

2. Suppose a gambler with initial capital k plays against an infinitely rich adversary and wins two units or loses one unit with probabilities p and q, respectively ($p+q=1$). What is the gambler's chance of being ruined?

3. Investigate the case of a simple random walk in one dimension, in which the walk starts from $x = k$; the probabilities of moving one step to the right or left are p and q, respectively; $x = a$ is an absorbing barrier and $x = 0$ is an elastic barrier (as defined at the end of Section 4.1).
What is the probability of absorption at $x = 0$?

4. Consider the one-dimensional random walk with absorbing barriers at $x = 0$ and $x = a$. Let $q_{kn}(\xi)$ be the probability that a walk starting at $x = k$ carries the particle at the nth step to the point $x = \xi$.

What are the difference equations and boundary conditions which would have to be solved in order to calculate $q_{kn}(\xi)$ explicitly?

5. Suppose we have a two-dimensional unrestricted symmetric random walk starting from the origin. Let r_n be the distance of the moving particle from the origin after the nth step. Prove that $E(r_n^2) = n$.

CHAPTER 5

Markov Chains

5.1 Introduction

In our previous discussion of recurrent events in Chapter 3 we started off by envisaging a sequence of independent trials, each involving for example the tossing of a coin. From consideration of the outcome of each individual trial we passed to more complex patterns such as the event "the accumulated numbers of heads and tails are equal". And in the formal definition of a recurrent event it was not necessary to assume that the basic trials were independent of one another; only that once the event in question had occurred the process started off again from scratch.

We now want to generalize this idea slightly so as to be able to take account of several possible outcomes at each stage, but shall introduce the restriction that the future probability behavior of the process is uniquely determined once the state of the system at the present stage is given. This restriction is the characteristic *Markov* property, and it is not so serious as might appear at first sight. It is equivalent, for example, to expressing the probability distribution of the size of a population tomorrow solely in terms of the size today, taking into account the effects of birth and death. In reality, of course, the population tomorrow is likely to be related not only to the size today, but also to the sizes on many previous days. Nevertheless, the proposed restriction frequently enables us to formulate models that are both mathematically tractable and useful as first approximations to a stochastic picture of reality. Moreover, as will appear in the sequel, there are several ways in which the more elaborate models can often be handled by methods appropriate to the restricted situation, such as, for example, concentrating attention on a series of elements having the Markov property.

For a more extensive account of the theory of Markov chains, see Feller (1957, Chapter 15); or the whole book by Kemeny and Snell (1960).

5.2   Notation and definitions

Before elaborating a mathematical discussion of the kind of process referred to in the last section, we must first introduce an adequate notation and definitions of the basic concepts.

Suppose we have a sequence of consecutive trials, numbered $n = 0, 1, 2, \cdots$. The outcome of the nth trial is represented by the random variable X_n, which we shall assume to be discrete and to take one of the values $j = 1, 2, \cdots$. The actual set of outcomes at any trial is a system of events $E_i, i = 1, 2, \cdots$, that are mutually exclusive and exhaustive: these are the *states of the system*, and they may be finite or infinite in number.

Thus in a simple series of coin-tossing trials, there are just two events: $E_1 \equiv$ heads, say, and $E_2 \equiv$ tails. For convenience we could label these events 1 and 2. On the other hand, in a series of random digits the events could be the actual numbers 0 to 9. And in a population of animals, the events might be the numbers of new births in each generation.

We next consider the probabilities of the various events. Let us write the absolute probability of outcome E_j at the nth trial as

$$P\{X_n = j\} = p_j^{(n)}, \tag{5.1}$$

so that the initial distribution is given by $p_j^{(0)}$.

If we have $X_{n-1} = i$ and $X_n = j$, we say that the system has made a *transition* of type $E_i \to E_j$ at the nth trial or step. Accordingly, we shall want to know the probabilities of the various transitions that may occur. If the trials are not independent, this means that in general we have to specify

$$P\{X_n = j | X_{n-1} = i, X_{n-2} = h, \cdots, X_0 = a\}. \tag{5.2}$$

With independent trials, the expressions in (5.1) and (5.2) are of course identical.

Now, as already indicated in the previous section, the Markov property means that the future behavior of a sequence of events is uniquely decided by a knowledge of the present state. In such a case the transition probabilities in (5.2) depend only on X_{n-1}, and *not* on the previous random variables. We can therefore define a *Markov chain* as a sequence of consecutive trials such that

$$P\{X_n = j | X_{n-1} = i, \cdots, X_0 = a\} = P\{X_n = j | X_{n-1} = i\}. \tag{5.3}$$

(It is possible to define Markov chains of a higher order, in which the expression on the right of (5.3) involves some fixed number of stages prior to the nth greater than one. We shall not, however, pursue this generalization here.)

Now an important class of chains defined by the property in (5.3) is that for which the transition probabilities are independent of n. We then have a *homogeneous Markov chain*, for which

$$P\{X_n = j \,|\, X_{n-1} = i\} = p_{ij}. \tag{5.4}$$

Note that the order of the subscripts in p_{ij} corresponds to the direction of the transition, i.e. $i \rightarrow j$. We must have, of course,

$$\sum_{j=1}^{\infty} p_{ij} = 1, \qquad i = 1, 2, \cdots \left.\begin{array}{c} \\ \end{array}\right\}$$

and
$$p_{ij} \geqslant 0 \qquad\qquad \tag{5.5}$$

since, for any fixed i, the p_{ij} will form a probability distribution.

In typical applications we are likely to be given the initial distribution and the transition probabilities, and we want to determine the probability distribution for each random variable X_n. In particular, we may be interested in the limiting distribution of X_n as $n \rightarrow \infty$, if this exists.

The transition probabilities are most conveniently handled in matrix form. Let us write, therefore, $\mathbf{P} = \{p_{ij}\}'$, i.e.

$$\mathbf{P} = \begin{bmatrix} p_{11} & p_{21} & p_{31} & \cdots \\ p_{12} & p_{22} & p_{32} & \cdots \\ p_{13} & p_{23} & p_{33} & \cdots \\ \vdots & \vdots & \vdots & \end{bmatrix}. \tag{5.6}$$

This is the *transition matrix*. It will be of finite or infinite order, depending on the number of states involved. The elements will all be non-negative, and in virtue of (5.5) the *columns* all sum to unity: a matrix with the latter property is often called a *stochastic matrix*. (We could have used the alternative definition $\mathbf{P} = \{p_{ij}\}$, instead of $\mathbf{P} = \{p_{ij}\}'$, but the latter transposed form is algebraically a little more convenient.)

In order to determine the absolute probabilities at any stage, we shall need the idea of n-step transition probabilities, where n can now be greater than unity. Let us in fact write

$$P\{X_{n+m} = j \,|\, X_m = i\} = p_{ij}^{(n)} \left.\begin{array}{c} \\ \end{array}\right\}$$

with
$$p_{ij}^{(1)} = p_{ij} \qquad\qquad \tag{5.7}$$

We have written the n-step transition probability $p_{ij}^{(n)}$ as being independent of m. This is in fact true for homogeneous chains, as we can easily show.

Suppose we write the absolute probabilities $p_j^{(n)}$ at the nth stage as the column vector $\mathbf{p}^{(n)}$, the initial distribution being $\mathbf{p}^{(0)}$. Now the distribution at the first stage is given by

$$p_j^{(1)} = \sum_{i=1}^{\infty} p_{ij} p_i^{(0)}, \qquad (5.8)$$

since the probability that the system is in state i initially is $p_i^{(0)}$, and the probability of a transition from i to j is p_{ij}, and we must sum over all transitions leading to state j. In matrix terms, we can express (5.8) as

$$\mathbf{p}^{(1)} = \mathbf{P}\mathbf{p}^{(0)}. \qquad (5.9)$$

Similarly,

$$\mathbf{p}^{(2)} = \mathbf{P}\mathbf{p}^{(1)} = \mathbf{P}^2 \mathbf{p}^{(0)}, \qquad (5.10)$$

and in general

$$\mathbf{p}^{(n)} = \mathbf{P}^n \mathbf{p}^{(0)}. \qquad (5.11)$$

We also have

$$\mathbf{p}^{(n+m)} = \mathbf{P}^{n+m} \mathbf{p}^{(0)}$$

$$= \mathbf{P}^n (\mathbf{P}^m \mathbf{p}^{(0)})$$

$$= \mathbf{P}^n \mathbf{p}^{(m)}. \qquad (5.12)$$

The matrix $\mathbf{P}^n$ therefore gives the required set of n-step transition probabilities $\{p_{ij}^{(n)}\}$, and the derivation shows that, for homogeneous chains at least, this is independent of m.

Equation (5.11) is of basic importance. It shows how to calculate the absolute probabilities at any stage in terms of the initial distribution $\mathbf{p}^{(0)}$ and the transition matrix $\mathbf{P}$. The main labor in actual calculations is of course the evaluation of the nth power of the matrix $\mathbf{P}$. In principle we could always proceed step by step. But in practice there is a considerable advantage in using special methods of calculating $\mathbf{P}^n$ more directly. One such method is described in detail in Section 5.5.

5.3 Classification of states

A good deal of the practical importance of Markov chain theory attaches to the fact that the states can be classified in a very distinctive manner according to certain basic properties of the system. An outline of the main classification, together with some of the more important theorems involved, is given below.

If the state E_j can be arrived at from the state E_i in a finite number of steps with non-zero probability, we say that it can be *reached*; i.e. there is a number $n > 0$ such that $p_{ij}^{(n)} > 0$. Thus in an unrestricted random walk, for example, any point on the line can be reached from any other point.

But if there is an absorbing barrier, no other state can be reached once the barrier has been arrived at.

If every state in a Markov chain can be reached from every other state, the chain is said to be *irreducible*. Alternatively, we may have a set of states C, which are closed in the sense that it is impossible to reach any state outside C from any state of C itself by one-step transitions. More precisely, $p_{ij} = 0$ if $E_i \subset C$ and $E_j \not\subset C$. In particular, if a single state E_k forms a closed set it is called an *absorbing state*, and $p_{kk} = 1$.

It follows from these definitions that in an irreducible Markov chain the set of all states constitutes a closed set, and no other set can be closed.

The importance of the concept of a closed set of states lies in the fact that we can envisage a sub-Markov chain defined on the closed set; and such a chain can be studied independently of the remaining states.

Example

In order to illustrate the items of classification introduced so far, let us consider a Markov chain discussed by Feller (1957, Chapter 14, Section 4) for which the transition matrix is given, at least in skeleton form, below. It is of course necessary to indicate only which elements are zero (shown by a dot) and which are positive (shown by a cross). Suppose we have

$$\mathbf{P} = \begin{bmatrix} \cdot & \cdot & \cdot & \times & \cdot & \cdot & \cdot & \cdot & \cdot \\ \cdot & \times & \cdot & \cdot & \cdot & \times & \times & \cdot & \cdot \\ \cdot & \times & \cdot & \cdot & \cdot & \cdot & \cdot & \times & \cdot \\ \times & \cdot & \cdot & \cdot & \cdot & \cdot & \cdot & \cdot & \times \\ \cdot & \times & \cdot & \cdot & \times & \cdot & \cdot & \cdot & \cdot \\ \cdot & \cdot & \cdot & \cdot & \cdot & \cdot & \times & \cdot & \cdot \\ \cdot & \cdot & \cdot & \cdot & \cdot & \cdot & \times & \cdot & \cdot \\ \cdot & \cdot & \times & \cdot & \cdot & \cdot & \cdot & \cdot & \cdot \\ \times & \times & \cdot & \cdot & \cdot & \cdot & \cdot & \cdot & \times \end{bmatrix} . \quad (5.13)$$

First, we examine the columns of the transition matrix for columns containing a single positive element in the leading diagonal. The fifth column is precisely of this type. It is clear that the actual entry must be $p_{55} = 1$, and that E_5 is accordingly an absorbing state.

Next, we look at columns with a single entry in some position other than on the leading diagonal. The third, fourth, sixth, and eighth columns are of this type, but we notice a reciprocal relation between the third and eighth columns, in that $p_{38} = 1 = p_{83}$. The states E_3 and E_8 evidently form a closed set, but nothing special can be said about E_4 and E_6.

Now from E_1 we can go to E_4 and E_9 in a single step; from E_4 we return to E_1; and from E_9 we can move to E_4 or remain in E_9. The three states E_1, E_4 and E_9 therefore constitute another closed set.

A clearer picture is obtained by suitably relabeling the states as follows:

$$E_1' = E_5; \quad E_2' = E_3; \quad E_3' = E_8; \quad E_4' = E_1; \quad E_5' = E_9;$$

$$E_6' = E_4; \quad E_7' = E_2; \quad E_8' = E_7; \quad E_9' = E_6.$$

The transition matrix now takes the form

$$\mathbf{Q} = \begin{bmatrix} \times & \cdot & \cdot & \cdot & \cdot & \cdot & \times & \cdot & \cdot \\ \cdot & \cdot & \times & \cdot & \cdot & \cdot & \times & \cdot & \cdot \\ \cdot & \times & \cdot & \cdot & \cdot & \cdot & \cdot & \cdot & \cdot \\ \cdot & \cdot & \cdot & \cdot & \cdot & \times & \cdot & \cdot & \cdot \\ \cdot & \cdot & \cdot & \times & \times & \cdot & \times & \cdot & \cdot \\ \cdot & \cdot & \cdot & \times & \times & \cdot & \cdot & \cdot & \cdot \\ \cdot & \cdot & \cdot & \cdot & \cdot & \cdot & \times & \times & \times \\ \cdot & \cdot & \cdot & \cdot & \cdot & \cdot & \cdot & \times & \cdot \\ \cdot & \cdot & \cdot & \cdot & \cdot & \cdot & \cdot & \times & \cdot \end{bmatrix}. \quad (5.14)$$

The closed sets (E_1'); (E_2', E_3') and (E_4', E_5', E_6') are now evident at a glance. Thus we could consider the corresponding stochastic matrices

$$[\times]; \quad \begin{bmatrix} \cdot & \times \\ \times & \cdot \end{bmatrix}; \quad \begin{bmatrix} \cdot & \cdot & \times \\ \times & \times & \cdot \\ \times & \times & \cdot \end{bmatrix};$$

independently of the rest of the array. Note that the last three states, E_7', E_8' and E_9' are not independent of the first six, although the latter *are* independent of the former.

Now it was indicated in Section 5.1 that we proposed to generalize our previous account of recurrent events in Chapter 3 to include situations with several possible outcomes at each stage. It should be immediately obvious that the definition of a Markov chain in the present chapter entails the following propositions. First, if we consider any arbitrary state E_j, and suppose that the system is initially in E_j, then "a return to E_j" is simply a recurrent event as discussed in Section 3.2. Secondly, if the system is initially in some other state E_i, "a passage to E_j" is a delayed recurrent event, as defined in Section 3.5. We can, therefore, make a direct application of the theory of recurrent events to deal with the present case of Markov chains.

Let us write $f_j^{(n)}$ for the probability that, starting from state E_j, the first return to E_j occurs at precisely the nth step. In the notation of Section 3.2, we have

$$p_{jj}^{(n)} = u_n, \quad f_j^{(n)} = f_n. \quad (5.15)$$

Formula (3.4) now takes the form

$$p_{jj}^{(n)} = \sum_{m=1}^{n} f_j^{(m)} p_{jj}^{(n-m)}, \quad n \geqslant 1, \tag{5.16}$$

where, conventionally, $p_{jj}^{(0)} = 1$. If we put (5.16) in the form

$$f_j^{(n)} = p_{jj}^{(n)} - \sum_{m=1}^{n-1} f_j^{(m)} p_{jj}^{(n-m)}, \quad n \geqslant 1, \tag{5.17}$$

then it is clear that we have a set of recurrence relations giving the recurrence time distribution $\{f_j^{(n)}\}$ in terms of the $p_{jj}^{(n)}$.

In particular, we can define the probability that the system returns *at least once* to E_j as f_j, where

$$f_j = \sum_{n=1}^{\infty} f_j^{(n)}. \tag{5.18}$$

We can therefore call the state E_j *persistent* if $f_j = 1$, with mean recurrence time

$$\mu_j = \sum_{n=1}^{\infty} n f_j^{(n)}, \tag{5.19}$$

and *transient* if $f_j < 1$, in which case we have $\mu_j = \infty$. From Theorem 3.1, it follows that a necessary and sufficient condition for E_j to be persistent is that $\sum p_{jj}^{(n)}$ should diverge, though this is often hard to apply.

If the state E_j is persistent, but $\mu_j = \infty$, we say that E_j is a *null state*.

We must also consider the possibility of a periodicity in the return to a given state. Thus the definition at the end of Section 3.2 can be applied, and we say that the state E_j is *periodic* with period $t > 1$, if E_j can occur only at stages numbered $t, 2t, 3t, \cdots$. When $t = 1$, we say that E_j is *aperiodic*.

If E_j is persistent, aperiodic, and not null, it is said to be *ergodic*, i.e. $f_j = 1, \mu_j < \infty, t = 1$.

Now let T be the set of transient states. The remaining persistent states can be divided into mutually disjoint closed sets, $C_1, C_2, \cdots$, such that from any state of one of these sets all states of this set, and no others, can be reached. Moreover, states in C_n may be reached from T, but not conversely. It can be shown that all states of a closed set C_n must have the same period, and so we can speak of the period of C_n.

An important result in the classification of Markov chains is that all states of an irreducible chain are of the same type: that is, they are all transient; all persistent and null; or all persistent and non-null. Moreover, in each case all states have the same period.

We now consider for a moment the previous illustration of a Markov chain whose transition matrix $\mathbf{Q}$ is given by (5.14). Dropping primes for convenience, we have E_1 as an absorbing state, and therefore persistent.

From E_2 the system must pass to E_3, and then back to E_2. The states E_2 and E_3 are thus persistent, and are periodic with period 2. We can also see that states E_4, E_5 and E_6 form a closed subset which is persistent and aperiodic. So far as the last three states are concerned, we first note that from E_7 each of the three closed sets can be reached, and on entering one of these sets the system stays there. There is thus a non-zero chance of no return, and E_7 is accordingly transient. Similarly, from E_9 we go to E_7, with no possible return to E_9. So the latter is also transient. Finally, from E_8 the system arrives sooner or later at E_7, never to return. Hence E_7, E_8, and E_9 are all transient states.

Having given the main classification required for individual states, based on the definitions of recurrent events introduced in Section 3.2, we next examine the delayed type of recurrent event which occurs when we consider a passage to E_j from some other state E_i. Let us write $f_{ij}^{(n)}$ for the probability that, starting from state E_i, the first passage to E_j occurs at precisely the nth step. In the notation of Section 3.2 we now have

$$p_{ij}^{(n)} = u_n, \qquad f_{ij}^{(n)} = f_n, \tag{5.20}$$

and the analog of equation (5.16) is

$$p_{ij}^{(n)} = \sum_{m=1}^{n} f_{ij}^{(m)} p_{ij}^{(n-m)}, \quad n \geqslant 1, \tag{5.21}$$

which may be put in the form

$$f_{ij}^{(n)} = p_{ij}^{(n)} - \sum_{m=1}^{n-1} f_{ij}^{(n)} p_{ij}^{(n-m)}, \quad n \geqslant 1, \tag{5.22}$$

corresponding to (5.17). The probability that, starting from E_i, the system ever reaches E_j, is f_{ij}, where

$$f_{ij} = \sum_{n=1}^{\infty} f_{ij}^{(n)}, \tag{5.23}$$

which is the appropriate extension of (5.18).

We are now in a position to state some results of fundamental importance to the theory of Markov chains. These are obtained directly from Theorems 3.2, 3.3, and 3.4, suitably interpreted in the present context.

Theorem 5.1

If E_j is persistent and aperiodic, with mean recurrence time μ_j, then

$$p_{jj}^{(n)} \to \mu_j^{-1} \quad \text{as} \quad n \to \infty, \tag{5.24}$$

and in particular, if E_j is also null, i.e. $\mu_j = \infty$,

$$p_{jj}^{(n)} \to 0 \quad \text{as} \quad n \to \infty.$$

Theorem 5.2

If E_j is persistent and periodic with period t, then

$$p_{jj}^{(nt)} \to t\mu_j^{-1} \quad \text{as} \quad n \to \infty. \tag{5.25}$$

Theorem 5.3

If E_j is persistent and aperiodic, then defining the probability that the system ever reaches E_j from E_i by f_{ij}, as given in (5.23), we have

$$p_{ij}^{(n)} \to \mu_j^{-1} f_{ij} \quad \text{as} \quad n \to \infty. \tag{5.26}$$

In particular, if E_j is also null, $\mu_j = \infty$, and

$$p_{ij}^{(n)} \to 0 \quad \text{as} \quad n \to \infty.$$

A useful corollary of the foregoing results is that it can be shown that in a *finite* Markov chain (i.e. a chain with only a finite number of states) there can be no null states, and it is impossible for all states to be transient.

5.4 Classification of chains

From a classification of the individual states of a Markov chain we pass to a classification of the chain as a whole, together with a number of important theorems about the behavior of certain types of chains.

We have already seen that in an irreducible chain all states belong to the same class, i.e. they are all transient; all persistent and null; or all persistent and non-null; moreover, all states have the same period.

Now, we call the chain *ergodic* if the absolute probability distributions $\mathbf{p}^{(n)}$ converge to a limiting distribution independently of the initial distribution $\mathbf{p}^{(0)}$, i.e. if

$$\lim_{n \to \infty} \mathbf{p}^{(n)} = \mathbf{p}. \tag{5.27}$$

And it can be proved that if all states of a chain are ergodic in the sense of the definition in Section 5.3, i.e. persistent, non-null, and aperiodic, then the chain itself is ergodic in the sense of the definition in equation (5.27) above.

The actual limiting distribution, if it exists, can of course be determined quite easily if the limit of the n-step transition $p_{jk}^{(n)}$ is known. In many practical cases the formation of the nth power of the transition matrix enables us to do just this quite explicitly (see Section 5.5 below).

Let us now refer back to the fundamental equation (5.11), i.e.

$$\mathbf{p}^{(n)} = \mathbf{P}^n \mathbf{p}^{(0)}.$$

It is clear that a solution of the equation

$$\mathbf{p} = \mathbf{Pp}, \tag{5.28}$$

with $\sum p_i = 1$ and $p_i \geqslant 0$, is a stationary distribution in the sense that if it is chosen for an initial distribution, all subsequent distributions $\mathbf{p}^{(n)}$ will also be identical with $\mathbf{p}$.

Another valuable theorem is that if a Markov chain is ergodic, then the limiting distribution is stationary; this is the only stationary distribution, and it is obtained by solving equation (5.28) above.

Two further theorems are worth quoting in respect of the existence of ergodicity:

Theorem 5.4

All states of a finite, aperiodic, irreducible Markov chain are ergodic, and hence the chain itself is ergodic.

Theorem 5.5

An irreducible and aperiodic Markov chain is ergodic if we can find a non-null solution of

$$\mathbf{x} = \mathbf{Px},$$

with $\sum |x_i| < \infty$ (this is Foster's theorem).

5.5 Evaluation of $\mathbf{P}^n$

As indicated at the end of Section 5.2, a problem of fundamental practical importance in the handling of Markov chains is the evaluation of the matrix $\mathbf{P}^n$, which gives the whole set of n-step transition probabilities. Because of its importance in actually constructing the required solution, it is worth giving the appropriate mathematical derivation in some detail.

Let us begin by supposing that we have a Markov chain with a finite number of states k, so that the transition matrix $\mathbf{P}$ is a $k \times k$ array. We first assume that the eigenvalues of $\mathbf{P}$, i.e. the roots of the characteristic equation

$$|\mathbf{P} - \lambda \mathbf{I}| = 0, \tag{5.29}$$

are all distinct.

Now the equations

$$\mathbf{Px} = \lambda \mathbf{x} \quad \text{and} \quad \mathbf{y}'\mathbf{P} = \lambda \mathbf{y}' \tag{5.30}$$

have solutions $\mathbf{x}$ and $\mathbf{y}$ other than zero, if and only if λ is an eigenvalue of $\mathbf{P}$. For each λ_j, therefore, we can find the corresponding left and right eigenvectors $\mathbf{x}_j$ and $\mathbf{y}'_j$.

We next define

$$H = [x_1, x_2, \cdots, x_k], \qquad K = [y_1, y_2, \cdots, y_k]. \tag{5.31}$$

It immediately follows from (5.30) that

$$PH = H\Lambda \quad \text{and} \quad K'P = \Lambda K', \tag{5.32}$$

where Λ is the diagonal matrix

$$\Lambda = \begin{bmatrix} \lambda_1 & 0 & \cdots & 0 \\ 0 & \lambda_2 & \cdots & 0 \\ \vdots & \vdots & & \vdots \\ 0 & 0 & \cdots & \lambda_k \end{bmatrix} \tag{5.33}$$

From (5.32) we thus have the alternative expressions for P given by

$$P = H\Lambda H^{-1} = (K')^{-1}\Lambda K'. \tag{5.34}$$

The right and left eigenvectors satisfy the relations

$$y_j'Px_i = y_j'(\lambda_i x_i) = (y_j'\lambda_j)x_i, \tag{5.35}$$

in virtue of the equations in (5.30). If the eigenvalues are distinct it immediately follows from (5.35) that

$$y_j'x_i = 0, \quad i \neq j. \tag{5.36}$$

Further, we can obviously choose the scales of x_j and y_j so that

$$y_j'x_j = 1. \tag{5.37}$$

Using (5.36) and (5.37), we see that

$$K'H = I, \tag{5.38}$$

and so (5.34) can be put in the form

$$P = H\Lambda K' = \sum_{j=1}^{k} \lambda_j A_j \left.\vphantom{\sum_{j=1}^{k}}\right\} \tag{5.39}$$

where

$$A_j = x_j y_j'$$

The matrices A_j constitute the *spectral set*, and they satisfy the relations

$$\left. \begin{aligned} A_i A_j &= 0, \quad i \neq j \\ &= A_j, \quad i = j \\ \sum_{j=1}^{k} A_j &= I \end{aligned} \right\} \tag{5.40}$$

and

The first two properties are easily verified by expressing the $\mathbf{A}_j$ in terms of the eigenvectors $\mathbf{x}_j$ and $\mathbf{y}_j$, and using (5.36) and (5.37). The last follows from

$$\sum \mathbf{A}_j = \mathbf{HK}' = \mathbf{HK}'\mathbf{HH}^{-1} = \mathbf{I}.$$

We can now use the expansion in (5.39) to write down the nth power of $\mathbf{P}$ immediately. Because of the relations (5.40) satisfied by the spectral set of matrices $\mathbf{A}_j$, we have

$$\mathbf{P}^n = \sum_{j=1}^{k} \lambda_j{}^n \mathbf{A}_j. \tag{5.41}$$

In practical calculations it is sometimes more convenient not to adjust the scales of $\mathbf{x}_j$ and $\mathbf{y}_j$ so as to make $\mathbf{y}_j'\mathbf{x}_j = 1$. If in fact

$$\mathbf{y}_j'\mathbf{x}_j = c_j, \tag{5.42}$$

we can put (5.41) in the form

$$\mathbf{P}^n = \sum_{j=1}^{k} c_j{}^{-1} \lambda_j{}^n \mathbf{A}_j, \tag{5.43}$$

where $\mathbf{A}_j$ is still defined as $\mathbf{x}_j \mathbf{y}_j'$.

The main labor in actually applying this technique lies first in calculating the eigenvalues, and secondly in constructing the eigenvectors. The element in any column of $\mathbf{P}$ must sum to unity where $\mathbf{P}$ is a stochastic matrix. We can therefore always choose one eigenvector to be $\mathbf{y}_1' = [1, 1, \cdots, 1]$, since

$$[1, 1, \cdots, 1]\mathbf{P} = [1, 1, \cdots, 1].$$

Accordingly, there is always one eigenvalue $\lambda_1 = 1$. And it can be shown that no other eigenvalue can be greater in absolute value.

It is a well-known result in matrix algebra that a square matrix $\mathbf{M} = \{m_{ij}\}$ satisfies

$$\mathbf{M} \text{ adj } \mathbf{M} = |\mathbf{M}|\mathbf{I}, \tag{5.44}$$

where the adjoint matrix adj $\mathbf{M}$ is simply the transpose of the matrix of cofactors, i.e. $\{M_{ji}\}$. If we now put

$$\mathbf{M} = \mathbf{P} - \lambda_j \mathbf{I}$$

in (5.44), we have

$$(\mathbf{P} - \lambda_j\mathbf{I})\text{adj}(\mathbf{P} - \hat{\lambda}_j\mathbf{I}) = |\mathbf{P} - \lambda_j\mathbf{I}|\mathbf{I}$$

$$= 0, \tag{5.45}$$

from (5.29). We can therefore choose any column of $\text{adj}(\mathbf{P} - \lambda_j\mathbf{I})$ as an eigenvector $\mathbf{x}_j$. Similarly $\mathbf{y}_j'$ can be derived from any row.

Alternatively, it is known that any matrix must satisfy its own characteristic equation. Since (5.29) can be factorized in the form

$$\prod_{j=1}^{k} (\lambda - \lambda_j) = 0, \tag{5.46}$$

we have

$$\prod_{j=1}^{k} (\mathbf{P} - \lambda_j \mathbf{I}) \equiv (\mathbf{P} - \lambda_j \mathbf{I}) \prod_{i \neq j} (\mathbf{P} - \lambda_i \mathbf{I}) = 0. \tag{5.47}$$

Thus any column of

$$\prod_{i \neq j} (\mathbf{P} - \lambda_i \mathbf{I})$$

could be taken for the eigenvector $\mathbf{x}_j$, and any row can be used for $\mathbf{y}_j'$.

When the matrix $\mathbf{P}$ has multiple eigenvalues the appropriate formula for $\mathbf{P}^n$ becomes rather more complicated. In general (see Frazer, Duncan, and Collar, 1946) we can write

$$\left. \begin{aligned} \mathbf{P}^n &= \sum_j \frac{1}{(r_j - 1)!} \left[\frac{d^{r_j - 1}}{d\lambda^{r_j - 1}} \frac{\lambda^n \, \text{adj}(\lambda \mathbf{I} - \mathbf{P})}{\psi_j(\lambda)} \right]_{\lambda = \lambda_j}, \\ \psi_j(\lambda) &= \prod_{i \neq j} (\lambda - \lambda_i)^{r_i} \end{aligned} \right\} \tag{5.48}$$

where

and r_j is the multiplicity of λ_j.

We have treated only the case of Markov chains with a finite number of states k. It is sometimes possible to extend the method to chains with a denumerably infinite number of states, though we shall not undertake a discussion of this topic here.

In handling a large transition matrix we may not only be forced to use numerical methods to calculate some, or all except $\lambda_1 = 1$, of the eigenvalues, but may also be involved in a lengthy evaluation of the eigenvectors. In certain cases, however, the derivation can be simplified by suitably partitioning the matrix. If, for example, there are two closed sets of states, and some additional transient states, we could renumber the states so that the transition matrix took the partitioned form

$$\mathbf{P} = \begin{bmatrix} \mathbf{P}_1 & \mathbf{0} & \mathbf{A}_1 \\ \mathbf{0} & \mathbf{P}_2 & \mathbf{B}_1 \\ \mathbf{0} & \mathbf{0} & \mathbf{C} \end{bmatrix}. \tag{5.49}$$

The nth power of this can be written as

$$\mathbf{P}^n = \begin{bmatrix} \mathbf{P}_1^n & \mathbf{0} & \mathbf{A}_n \\ \mathbf{0} & \mathbf{P}_2^n & \mathbf{B}_n \\ \mathbf{0} & \mathbf{0} & \mathbf{C}^n \end{bmatrix}, \tag{5.50}$$

where the matrices $\mathbf{A}_n$ and $\mathbf{B}_n$ are *not* nth powers.

The matrices $\mathbf{P}_1{}^n$, $\mathbf{P}_2{}^n$ and $\mathbf{C}^n$, can be found separately by the previous spectral resolution method, and it may well be that these matrices are reasonably small and relatively easy to handle.

Elements such as $\mathbf{A}_n$ are a little more involved, but even here we need not manipulate the whole matrix. For it is easily seen that

$$\left[\begin{array}{cc} \mathbf{P}_1 & \mathbf{A}_1 \\ \mathbf{0} & \mathbf{C} \end{array} \right]^n = \left[\begin{array}{cc} \mathbf{P}_1^n & \mathbf{A}_n \\ \mathbf{0} & \mathbf{C}^n \end{array} \right], \qquad (5.51)$$

so that $\mathbf{A}_n$ can be calculated by finding the nth power of the "reduced" matrix on the left of (5.51). The last element $\mathbf{B}_n$ can also be found in a similar way by omitting the first row and column from $\mathbf{P}$ in (5.49).

5.6 Illustrations

Although the spectral resolution method of finding the nth power of a matrix described in the previous section may seem rather involved, it can often be applied in practice without too much difficulty. Let us examine two illustrations of the technique, one very simple, the other rather more complicated. These problems are investigated by Feller (1957), but treated by a different method.

A learning model

Suppose we consider the problem of an individual learning to make the correct response to a certain stimulus (e.g. Bush and Mosteller, 1951). Let E_1 be a correct response and E_2 an incorrect one. If we can assume that, in a series of successive trials, the response at any trial depends only on the response at the previous trial, then the series can be represented by a Markov chain with transition matrix

$$\mathbf{P} = \left[\begin{array}{cc} 1-p & \alpha \\ p & 1-\alpha \end{array} \right], \qquad (5.52)$$

where p is the chance of an incorrect result following a correct one, and α is the chance of a correct result following an incorrect one. If $p = 1 - \alpha$, each response is independent of the previous response; if $\frac{1}{2} < p, \alpha < 1$, there is a tendency to oscillate from one response to the other; and if $p > \alpha$, there is some kind of preference for correct responses. Thus, although this model is an extremely simple one, it does provide a certain choice of alternative interpretations.

It is in fact not too difficult in this case to form a few successive powers of $\mathbf{P}$ by direct matrix multiplication, and then to spot the general form of $\mathbf{P}^n$, which can then be verified by induction. To apply the standard method of Section 5.5, we must first solve the characteristic equation

$$\begin{vmatrix} 1 - p - \lambda & \alpha \\ p & 1 - \alpha - \lambda \end{vmatrix} = 0. \tag{5.53}$$

This multiplies out to give

$$\lambda^2 - (2 - \alpha - p)\lambda + (1 - \alpha - p) = 0, \tag{5.54}$$

for which the required roots are

$$\lambda = 1, \quad 1 - \alpha - p. \tag{5.55}$$

It is an easy matter to calculate adj$(\mathbf{P} - \lambda\mathbf{I})$ for both eigenvalues. Thus, when $\lambda = 1$, we have

$$\mathbf{P} - \mathbf{I} = \begin{bmatrix} -p & \alpha \\ p & -\alpha \end{bmatrix},$$

and therefore

$$\text{adj}(\mathbf{P} - \mathbf{I}) = \begin{bmatrix} -\alpha & -\alpha \\ -p & -p \end{bmatrix}. \tag{5.56}$$

Similarly, for the root $\lambda = 1 - \alpha - p$, we find

$$\mathbf{P} - (1 - \alpha - p)\mathbf{I} = \begin{bmatrix} \alpha & \alpha \\ p & p \end{bmatrix},$$

and

$$\text{adj}(\mathbf{P} - (1 - \alpha - p)\mathbf{I}) = \begin{bmatrix} p & -\alpha \\ -p & \alpha \end{bmatrix}. \tag{5.57}$$

From (5.53) and (5.54) we can choose eigenvectors

$$\left. \begin{array}{cc} \mathbf{x}_1 = \begin{bmatrix} \alpha \\ p \end{bmatrix}, & \mathbf{x}_2 = \begin{bmatrix} 1 \\ -1 \end{bmatrix} \\ \mathbf{y}_1' = [1, 1], & \mathbf{y}_2' = [p, -\alpha] \end{array} \right\} \tag{5.58}$$

Here we have

$$\left. \begin{array}{l} c_1 = \mathbf{y}_1'\mathbf{x}_1 = \alpha + p \\ c_2 = \mathbf{y}_2'\mathbf{x}_2 = \alpha + p \end{array} \right\}, \tag{5.59}$$

and

$$\mathbf{x}_1\mathbf{y}_1' = \begin{bmatrix} \alpha & \alpha \\ p & p \end{bmatrix}, \quad \mathbf{x}_2\mathbf{y}_2' = \begin{bmatrix} p & -\alpha \\ -p & \alpha \end{bmatrix}. \tag{5.60}$$

Thus

$$\mathbf{P}^n = \frac{(1)^n}{\alpha + p} \begin{bmatrix} \alpha & \alpha \\ p & p \end{bmatrix} + \frac{(1 - \alpha - p)^n}{\alpha + p} \begin{bmatrix} p & -\alpha \\ -p & \alpha \end{bmatrix}. \tag{5.61}$$

If the first response were correct, i.e.

$$\mathbf{p}^{(0)} = \begin{bmatrix} 1 \\ 0 \end{bmatrix},$$

the chance of a correct response at stage n would be the first element of $\mathbf{P}^n \mathbf{p}^{(0)}$, namely

$$\frac{\alpha}{\alpha + p} + \frac{p}{\alpha + p}(1 - \alpha - p)^n. \tag{5.62}$$

If $\alpha = p = 1$, there is oscillation from one state to the other, and no limiting probabilities exist. But if $\alpha, p < 1$, then $|1 - \alpha - p| < 1$ and the factor multiplying the second matrix on the right of (5.62) tends to zero. The limiting probability of success is thus $\alpha/(\alpha + p)$. It may be noted that this is less than unity unless $p = 0$, i.e. unless we can inhibit the tendency to revert to an incorrect response.

An inbreeding problem

Let us now look at a rather more involved application to a problem of inbreeding in genetics. Suppose we consider the behavior of an organism at a single locus, where the alternative genes are A and a. There are thus three possible genotypes, AA, Aa and aa. Suppose we mate two individuals and select, for the next generation, two of their offspring of opposite sexes. This procedure is then continued. It is in fact a schedule of brother-sister mating, and can be regarded as a Markov chain whose states are the six types of mating given by

$$E_1 = AA \times AA; \qquad E_2 = AA \times Aa; \qquad E_3 = Aa \times Aa;$$

$$E_4 = Aa \times aa; \qquad E_5 = aa \times aa; \qquad E_6 = AA \times aa.$$

The usual rules of Mendelian inheritance show that the transition matrix is

$$\mathbf{P} = \begin{bmatrix} 1 & \frac{1}{4} & \frac{1}{16} & 0 & 0 & 0 \\ 0 & \frac{1}{2} & \frac{1}{4} & 0 & 0 & 0 \\ 0 & \frac{1}{4} & \frac{1}{4} & \frac{1}{4} & 0 & 1 \\ 0 & 0 & \frac{1}{4} & \frac{1}{2} & 0 & 0 \\ 0 & 0 & \frac{1}{16} & \frac{1}{4} & 1 & 0 \\ 0 & 0 & \frac{1}{8} & 0 & 0 & 0 \end{bmatrix}. \tag{5.63}$$

We now have the main ingredients required for calculating the progress of the inbreeding program, starting from any chosen mating. In practice E_6 would be quite a likely initial mating, as it corresponds to an outcross between two "pure" lines.

It can be seen from the form of the transition matrix that E_1 and E_5 are closed sets, each of these states being absorbing; while the remaining states E_2, E_3, E_4, and E_6 are transient. A breeder might want to know the probability of absorption in E_1 and E_5, and the expected duration of the process, if the initial state were E_2, E_3, E_4, or E_6. Although in principle we require $\mathbf{P}^n$, we can obtain all the quantities we need, and incidentally avoid the complication of the repeated root $\lambda = 1$, by working with the reduced matrix $\mathbf{T}$ obtained by deleting the first and fifth rows and columns from $\mathbf{P}$, i.e.,

$$\mathbf{T} = \begin{bmatrix} \tfrac{1}{2} & \tfrac{1}{4} & 0 & 0 \\ \tfrac{1}{4} & \tfrac{1}{4} & \tfrac{1}{4} & 1 \\ 0 & \tfrac{1}{4} & \tfrac{1}{2} & 0 \\ 0 & \tfrac{1}{8} & 0 & 0 \end{bmatrix}. \tag{5.64}$$

Note that $\mathbf{T}$ is not a stochastic matrix. Its eigenvalues are found by solving the equation

$$\begin{vmatrix} \tfrac{1}{2} - \lambda & \tfrac{1}{4} & 0 & 0 \\ \tfrac{1}{4} & \tfrac{1}{4} - \lambda & \tfrac{1}{4} & 1 \\ 0 & \tfrac{1}{4} & \tfrac{1}{2} - \lambda & 0 \\ 0 & \tfrac{1}{8} & 0 & -\lambda \end{vmatrix} = 0. \tag{5.65}$$

On evaluating the determinant in (5.65), we obtain

$$(1 - 2\lambda)(1 - 4\lambda)(4\lambda^2 - 2\lambda - 1) = 0, \tag{5.66}$$

so that the eigenvalues are

$$\lambda_1 = \tfrac{1}{2}, \qquad \lambda_2 = \tfrac{1}{4}, \qquad \lambda_3 = \tfrac{1}{4}(1 + 5^{\frac{1}{2}}), \qquad \lambda_4 = \tfrac{1}{4}(1 - 5^{\frac{1}{2}}). \tag{5.67}$$

The solutions of the equations

$$\mathbf{T}\mathbf{x} = \lambda\mathbf{x}$$

can be found fairly easily by direct calculation in a straightforward way. They are, in order (apart from arbitrary multipliers),

$$[\mathbf{x}_1, \mathbf{x}_2, \mathbf{x}_3, \mathbf{x}_4] = \begin{bmatrix} 1 & 1 & 1 & 1 \\ 0 & -1 & -4\lambda_4 & -4\lambda_3 \\ -1 & 1 & 1 & 1 \\ 0 & -\tfrac{1}{2} & 2\lambda_4^2 & 2\lambda_3^2 \end{bmatrix}, \tag{5.68}$$

$$\begin{bmatrix} \mathbf{y}_1' \\ \mathbf{y}_2' \\ \mathbf{y}_3' \\ \mathbf{y}_4' \end{bmatrix} = \begin{bmatrix} 1 & 0 & -1 & 0 \\ 1 & -1 & 1 & -4 \\ 1 & -4\lambda_4 & 1 & 16\lambda_4^2 \\ 1 & -4\lambda_3 & 1 & 16\lambda_3^2 \end{bmatrix}. \tag{5.69}$$

The divisors c_i are given by

$$\left.\begin{array}{l} c_1 = \mathbf{y}_1'\mathbf{x}_1 = 2 \\[4pt] c_2 = \mathbf{y}_2'\mathbf{x}_2 = 5 \\[4pt] c_3 = \mathbf{y}_3'\mathbf{x}_3 = 8(1 + \lambda_4)^2 \\[4pt] c_4 = \mathbf{y}_4'\mathbf{x}_4 = 8(1 + \lambda_3)^2 \end{array}\right\}. \qquad (5.70)$$

We can therefore write

$$\mathbf{T}^n = \sum_{i=1}^{4} \lambda_i^n c_i^{-1}(\mathbf{x}_i\mathbf{y}_i'), \qquad (5.71)$$

where the λ_i are given by (5.67) and the c_i by (5.70); and the spectral matrices $\mathbf{x}_i\mathbf{y}_i'$ are formed from the vectors appearing in (5.68) and (5.69). The latter is numerically quite straightforward and is left as an exercise to the reader.

It remains to be shown how the actual probabilities of absorption at any step are calculated from (5.71). The latter formula, when developed in full, gives the n-step transition probabilities $p_{ij}^{(n)}$, for $i, j = 2, 3, 4, 6$.

Now absorption in E_1, say, at the nth step, starting in state E_i, necessarily implies that at step $n - 1$ the system enters E_2 or E_3 and passes to E_1 at the next step. This can be seen from inspection of the full transition matrix in (5.63). The appropriate probability is therefore

$$\tfrac{1}{4}p_{i2}^{(n-1)} + \tfrac{1}{16}p_{i3}^{(n-1)}, \qquad (5.72)$$

and the components $p_{i2}^{(n-1)}$ and $p_{i3}^{(n-1)}$ can be read directly from the developed form of (5.71) with $n - 1$ written for n.

In some problems it is possible to take a short cut and avoid explicit calculation of the eigenvectors, a process which is usually fairly tedious. For instance, suppose that in the above illustration we were content merely to calculate the expected proportion of heterozygous individuals at each stage. If, at stage n, this were q_n, we should have

$$q_n = \tfrac{1}{2}p_2^{(n)} + p_3^{(n)} + \tfrac{1}{2}p_4^{(n)}, \qquad (5.73)$$

since only the states E_2, E_3, and E_4 involve heterozygotes. Now if we were to evaluate the absolute probabilities $p_j^{(n)}$ in (5.73), we should clearly find that they could be written as appropriate linear functions of the nth power of the four eigenvalues in (5.67). Accordingly, q_n can be written in the form

$$q_n = a(\tfrac{1}{2})^n + b(\tfrac{1}{4})^n + c\left(\frac{1 + 5^{\frac{1}{2}}}{4}\right)^n + d\left(\frac{1 - 5^{\frac{1}{2}}}{4}\right)^n, \qquad (5.74)$$

where a, b, c, and d are suitable constants.

Now from any given initial state we can quite easily calculate q_0, q_1, q_2, and q_3, in succession, and this then provides us with 4 linear equations in the 4 constants.

For a further discussion of inbreeding applications see Fisher (1949).

PROBLEMS FOR SOLUTION

1. Show that a simple one-dimensional random walk, with chances p and q of one step to the right or left and absorbing barriers at $x = 0$ and $x = a$, can be represented by a Markov chain in which the point $x = j$ corresponds to the state E_j.

What is the transition matrix $\mathbf{P}$?

2. Consider a series of independent trials, at each of which there is success or failure with probabilities p and $q = 1 - p$, respectively. Let E_0 occur at the nth trial if the result is a failure, and E_k $(k = 1, 2, \cdots, n)$ if the result is a success with the previous failure at trial $n - k$, i.e. if there has been a sequence of k successes.

Write down the transition matrix $\mathbf{P}$.

3. Classify the Markov chains with the following transition matrices:

(i) $\begin{bmatrix} 0 & \frac{1}{2} & \frac{1}{2} \\ \frac{1}{2} & 0 & \frac{1}{2} \\ \frac{1}{2} & \frac{1}{2} & 0 \end{bmatrix}$ (ii) $\begin{bmatrix} 0 & 0 & \frac{1}{2} & 0 \\ 0 & 0 & \frac{1}{2} & 0 \\ 0 & 0 & 0 & 1 \\ 1 & 1 & 0 & 0 \end{bmatrix}$

(iii) $\begin{bmatrix} \frac{1}{2} & \frac{1}{2} & 0 & 0 & \frac{1}{4} \\ \frac{1}{2} & \frac{1}{2} & 0 & 0 & \frac{1}{4} \\ 0 & 0 & \frac{1}{2} & \frac{1}{2} & 0 \\ 0 & 0 & \frac{1}{2} & \frac{1}{2} & 0 \\ 0 & 0 & 0 & 0 & \frac{1}{2} \end{bmatrix}$

4. Suppose that $f_j^{(n)}$ and $p_{jj}^{(n)}$ are defined as in Sections 5.2 and 5.3 above. Let

$$F_j(x) = \sum_{n=1}^{\infty} f_j^{(n)} x^n, \quad P_j(x) = \sum_{n=1}^{\infty} p_{jj}^{(n)} x^n.$$

Prove that for $|x| < 1$ we have

$$F_j(x) = \frac{P_j(x)}{1 + P_j(x)}.$$

5. Consider a one-dimensional random walk with chances p and q of one step to the right or left, and reflecting barriers at $x = 0$ and $x = a$. If this is to be represented by a Markov chain, what is the transition matrix $\mathbf{P}$?

Obtain the stationary distribution.

6. Investigate the Markov chain with transition matrix

$$\mathbf{P} = \begin{bmatrix} p & q \\ q & p \end{bmatrix},$$

and initial distribution $\mathbf{p}^{(0)} = \begin{bmatrix} \alpha \\ \beta \end{bmatrix}$, where $\alpha + \beta = 1$. Work through the method

of Section 5.5 to obtain the n-step transition probabilities, the absolute distribution $\mathbf{p}^{(n)}$, and the limiting distribution $\mathbf{p} = \lim_{n \to \infty} \mathbf{p}^{(n)}$.

CHAPTER 6

Discrete Branching Processes

6.1 Introduction

We have seen in Chapter 5 how to handle processes that involve integral-valued variables and have the characteristic Markov property. It sometimes happens, however, that, although we can formulate the basic model of a certain stochastic process in terms of a suitable Markov chain, the transition probabilities are difficult to specify, and the methods of the last chapter are not readily applied. This happens, for example, if we have a population of individuals that reproduce in discrete generations at constant intervals. Typically, we can suppose that we start with one individual at the initial, or zero'th, generation, and that at each stage each individual has probability p_k of giving rise to k new individuals for the next generation. We could interpret $k = 0$ as corresponding to the death of the individual concerned; $k = 1$ as meaning the continued existence of the individual; and higher values of k as indicating the appearance of new individuals. The development of the whole population could therefore be pictured as a kind of family tree, branching out from the initial individual. This is obviously a special type of Markov chain, in so far as we specify that the individuals in any generation reproduce independently of past generations, though the appropriate transition matrix would be complicated to write down explicitly. A typical question asked of such a model is "what is the distribution of the total size of the population after n generations?" This problem can be investigated by the appropriate generating-function technique without recourse to the development of an n-step transition matrix.

In physics such discrete *branching processes* obviously occur in the context of nuclear chain reactions. In biology we may be concerned with problems such as following the future development of a mutant gene: we should be interested to calculate the numbers of mutant individuals likely to be present at any future generation; and in particular we should want to

58

know the probability of the mutant becoming extinct. A rather similar problem, which has been at least of historical importance, is the question of the survival of family names. Here only male descendants are taken into account, and p_j is the chance that any newborn boy will himself father j boys in the next generation. Of course p_j might change from generation to generation with secular variations in fertility, and statistical independence might be vitiated by a failure of the population to breed at random. Such complications can be ignored in elementary discussions, though it would be possible to take them into account.

In the present chapter we shall confine our attention to branching processes that have a fixed generation-time, so that they are really a special type of Markov chain. It is clear, however, that we should have a more general type of branching process if the generation-time were allowed to have a continuous distribution. This point will be taken up again later (see Section 8.9) when we discuss processes in continuous time. We shall also assume that all the individuals of a given generation reproduce independently of one another; this makes the process *multiplicative*.

6.2 Basic theory

It is convenient to begin by assuming that we start off with just one individual, i.e. $X_0 = 1$. This individual gives rise to j individuals in the next generation with probability p_j, where $j = 0, 1, 2, \cdots$. Thus the random variable X_1 for the population size at generation number "1" has the probability distribution given by $\{p_j\}$. Let the associated probability-generating function be

$$P(x) = \sum_{j=0}^{\infty} p_j x^j. \tag{6.1}$$

Now the second generation consists of the descendants of the first generation, and we suppose that each individual of the latter produces j offspring with probability p_j, independently of every other individual. The random variable X_2 for the size of the second generation is thus the sum of X_1 mutually independent random variables, each having the probability-generating function $P(x)$. The random variable X_2 therefore has a compound distribution in the sense of Section 2.4, and its probability-generating function $P_2(x)$ is given by

$$P_2(x) = P(P(x)) \tag{6.2}$$

using formula (2.27).

In general, it is obvious that we can write the probability-generating function $P_n(x)$ for the distribution of population size at the nth generation as

$$P_n(x) = P_{n-1}(P(x)), \quad n > 1, \tag{6.3}$$

since each individual at the $(n-1)$th generation produces offspring with probability-generating function $P(x)$.

Alternatively, each individual of the first generation gives rise to a number of descendants $n-1$ generations later having generating function $P_{n-1}(x)$. This is the nth generation, counting from the beginning. Hence the generating function $P_n(x)$ for this generation can also be written in the dual form

$$P_n(x) = P(P_{n-1}(x)), \quad n > 1. \tag{6.4}$$

When $n = 1$, we have of course simply

$$P_1(x) = P(x), \tag{6.5}$$

and when $n = 0$,

$$P_0(x) = x. \tag{6.6}$$

Thus formulas (6.3) and (6.4) apply to the case $n = 1$ also. Collecting the results together we have

$$P_0(x) = x; \qquad P_n(x) = P_{n-1}(P(x)) = P(P_{n-1}(x)), \quad n \geqslant 1. \tag{6.7}$$

If we started with a individuals, each of them would independently give rise to a branching process of the above type. The probability-generating function for the nth generation would therefore be $\{P_n(x)\}^a$. In terms of Markov chain theory, the n-step transition probability $p_{ij}^{(n)}$ is given by the coefficient of x^j in $\{P_n(x)\}^i$, i.e. the probability that i individuals give rise to j individuals in n generations.

Chance of extinction

Suppose we consider a process starting with a single individual. The chance that the population is extinct at the nth generation is $p_0^{(n)}$. This is the chance that extinction actually occurs on or before the nth step, so it is clear that $p_0^{(n)}$ must increase with n. If $p_0 = 0$, no extinction is possible since the population will always contain some individuals. We can, therefore, assume that $0 < p_0 \leqslant 1$.

Now the sequence $\{p_0^{(n)}\}$ must have a limit ζ since it is bounded and increasing. And since

$$p_0^{(n)} = P_n(0) = P(P_{n-1}(0)) = P(p_0^{(n-1)}),$$

we see that ζ satisfies

$$\zeta = P(\zeta). \tag{6.8}$$

Further, let x_0 be any arbitrary root of $x = P(x)$. Then, since $P(x)$ contains only positive terms and must increase in the interval $0 < x < 1$, we have

$$p_0^{(1)} = P(0) < P(x_0) = x_0,$$

and $\qquad p_0^{(2)} = P(p_0^{(1)}) < P(x_0) = x_0, \quad$ etc.

Hence $p_0^{(n)} < x_0$, by induction, and so ζ must be the *smallest positive root* of equation (6.8).

Next, we consider the behavior of the graph of $y = P(x)$ in the interval $(0, 1)$. The derivative $P'(x)$ contains only positive terms and must increase in the interval $0 < x < 1$. The curve $y = P(x)$ is thus convex and can intersect the line $y = x$ in two points at most. One of those must be $(1, 1)$, and so there is at most one other positive root ζ of the equation $x = P(x)$.

By considering the function $P'(x)$, it is easy to see that a necessary and sufficient condition for the existence of $\zeta < 1$ is that $P'(1) > 1$. For in this case we have, for diminishing x, the curve $y = P(x)$ falling below the line $y = x$, but eventually crossing it to reach $(0, y_0)$, where $y_0 = P(0) > 0$. If, on the other hand, $P'(1) \leqslant 1$, the curve $y = P(x)$ lies above $y = x$, and there is no root $\zeta < 1$. But $P'(1)$ is the mean value m of the distribution $\{p_j\}$. We thus have the result:

If m, the mean number of descendants of any individual in the next generation, is greater than unity, then a unique *root* $\zeta < 1$ of (6.8) exists and is the chance of the population becoming extinct after a finite number of generations. But if $m \leqslant 1$, the chance of extinction tends to unity.

The quantity $1 - \zeta$, in the case $m > 1$, is the chance of an infinitely prolonged process. In practice, we usually find that $p_0^{(n)}$ converges to ζ fairly quickly. This means that populations which are going to become extinct are likely to do so after only a few generations. We can thus interpret ζ as the chance of rapid extinction.

If we start with a individuals instead of only one, the chance that all descendant lines become extinct is ζ^a. The chance of at least one being successful is $1 - \zeta^a$. If a is large, $1 - \zeta^a$ will be near unity, even if ζ is relatively large. Thus, when a is large, as in a nuclear chain reaction, the condition $m \leqslant 1$ means that the process stops with probability one after a finite number of generations; while if $m > 1$, the chance of an "explosion" is near unity.

Calculation of cumulants

In spite of the iterative nature of the probability-generating functions involved in the above discussion, it is comparatively easy to calculate the mean and variance, etc., of the population size at each generation. In fact we can obtain very neat generalizations of the iterative formulas in (6.7) in terms of cumulant-generating functions. Writing as usual $M(\theta) = P(e^\theta)$, and $K(\theta) = \log M(\theta)$, we can develop (6.7) as follows. We have

$$M_n(\theta) = P_n(e^\theta)$$
$$= (P_{n-1}P(e^\theta)), \quad \text{using (6,7)},$$
$$= P_{n-1}(M(\theta))$$
$$= M_{n-1}(\log M(\theta))$$
$$= M_{n-1}(K(\theta)). \tag{6.9}$$

Taking logs of both sides of this equation gives immediately

$$K_n(\theta) = K_{n-1}(K(\theta)), \tag{6.10}$$

and a similar argument, using the alternative form $P_n(x) = P(P_{n-1}(x))$ in the second line, leads to the dual equation

$$K_n(\theta) = K(K_{n-1}(\theta)). \tag{6.11}$$

Differentiating (6.10) with respect to θ gives

$$K'_n(\theta) = K'_{n-1}(K(\theta))K'(\theta).$$

Putting $\theta = 0$, and writing the mean of the nth generation as m_n, now yields

$$m_n = m_{n-1}m. \tag{6.12}$$

Hence, applying (6.12) recursively, we find

$$m_n = m^n. \tag{6.13}$$

If we now differentiate (6.10) twice with respect to θ we obtain

$$K''_n(\theta) = K''_{n-1}(K(\theta))\{K'(\theta)\}^2 + K'_{n-1}(K(\theta))K''(\theta).$$

Writing $\theta = 0$, and using σ_n^2 for the variance at the nth generation, gives

$$\sigma_n^2 = m^2\sigma_{n-1}^2 + m_{n-1}\sigma^2$$
$$= m^2\sigma_{n-1}^2 + m^{n-1}\sigma^2. \tag{6.14}$$

Applying (6.14) recursively quickly leads to

$$\sigma_n^2 = (m^{n-1} + m^n + \cdots + m^{2n-2})\sigma^2$$
$$= \frac{m^{n-1}(m^n - 1)}{m - 1}\sigma^2, \qquad m \neq 1. \tag{6.15}$$

Similar arguments can be used to obtain higher cumulants. If $m = 1$, we obtain from the line before (6.15)

$$\sigma_n^2 = n\sigma^2, \quad m = 1. \tag{6.16}$$

Now the mean size of the nth generation, starting with a single individual is given by (6.13) and is m^n. Thus if $m > 1$ we expect a geometrical increase in population size. More specifically, we can investigate the behavior of the whole probability-generating function $P_n(x)$. We have $P(0) = p_0 > 0$, and as $P(x)$ is an increasing function of x, then $P(x) > x$ for $x < \zeta$. Therefore we can write $P_2(x) = P(P(x)) > P(x)$. Similarly, $P_3(x) = P(P_2(x)) > P_2(x)$ etc. Clearly $P_n(x)$ is an increasing sequence, and since $P_n(x) = P(P_{n-1}(x))$, it follows that $P_n(x) \to \zeta$. All the coefficients of $x, x^2, x^3, \cdots$ in $P_n(x)$ must therefore tend to zero as $n \to \infty$. We can interpret this as meaning that after a large number of generations the chance of the population being extinct is near ζ, while the chance of the number of individuals exceeding any arbitrary number is near $1 - \zeta$. The population is, accordingly, likely to be extinct or very large, with only a very small chance of being moderate in size.

6.3 Illustration

As an elementary illustration of the practical consequences of the previous theory let us consider a large and approximately constant population of individuals, in which a new mutant type appears in N of them. We suppose that both the normal and mutant individuals reproduce according to a discrete branching process, the normals producing only normals and the mutants producing only mutants. Of course, an accurate treatment of a higher organism with sexual reproduction would really require a more sophisticated model, since the genes occur in pairs. But the present simplified discussion may be taken to give some insight into the mechanism of spread of new types.

If m, the mean number of mutant invididuals produced by any one mutant, is less than or equal to unity, then the descendant line from any mutant is certain to die out sooner or later.

Next suppose that the mutant type has a slight reproductive advantage over the normal type, that is, we can write

$$m = 1 + \varepsilon, \qquad (6.17)$$

where ε is small and positive. We now ask about the chance of the mutant type becoming extinct, or, alternatively, the chance that it will establish itself. We require the smallest positive root ζ (between 0 and 1) of the equation

$$x = P(x), \qquad (6.18)$$

where $P(x)$ is the probability-generating function of the number of progeny produced by a single mutant. Given the precise form of the function $P(x)$, we could compute the required root numerically. More generally, let us see if we can find an approximation involving only the mean and variance, m and σ^2.

If we put $x = e^\theta$ in (6.18) we obtain

$$e^\theta = P(e^\theta) = M(\theta), \tag{6.19}$$

where $M(\theta)$ is the moment-generating function of the number of progeny produced by a single mutant. Taking logarithms of both sides of (6.19) yields

$$\theta = \log M(\theta)$$
$$= K(\theta)$$
$$= m\theta + \tfrac{1}{2}\sigma^2\theta^2 + \cdots, \tag{6.20}$$

where $K(\theta)$ is the cumulant-generating function. Ignoring terms on the right of (6.20) involving powers of θ higher than θ^2, gives an equation for θ, with root

$$\theta_0 \sim 2(1 - m)/\sigma^2. \tag{6.21}$$

From (6.21) we have

$$\zeta = e^{\theta_0} \doteq e^{2(1-m)/\sigma^2} = e^{-2\varepsilon/\sigma^2}. \tag{6.22}$$

The chance of extinction of all N mutants is

$$\zeta^N \doteq e^{-2N\varepsilon/\sigma^2}. \tag{6.23}$$

To take a numerical example, suppose that the distribution of offspring is Poisson in form with $m = \sigma^2 = 1 + \varepsilon$, with $\varepsilon = 0.01$. Then

$$\zeta^N \doteq e^{-0.0198N} = (0.9804)^N. \tag{6.24}$$

Thus if $N \geqslant 250$, the chances against extinction would be better than 100 to 1. Of course, this refers to the situation when there is initially an appreciable number of new types of individual having a small reproductive advantage.

An extension of these ideas can be made to the case where mutation occurs as a rare but repeated event. The chance of any single mutant's descendant line becoming extinct is ζ. But sooner or later there will have been a sufficient number of mutations for there to be a large probability of at least one of them establishing itself.

PROBLEMS FOR SOLUTION

1. By choosing a suitable generating function equation prove that

$$\sigma_n^2 = m\sigma_{n-1}^2 + m^{2(n-1)}\sigma^2.$$

Hence derive the result in equation (6.15).

2. Suppose that the bivariate generating function of the random variables X_j and X_k is given by

$$P(x, y) = E(x^r y^s) = \sum_{r,s} p_{rs} x^r y^s,$$

where

$$p_{rs} = P\{X_j = r, X_k = s\}.$$

Show that, for $k > j$,

$$P(x, y) = P_j(x P_{k-j}(y)).$$

3. Use the result of Problem 2 to obtain an expression for $\text{cov}(X_j, X_k)$.

CHAPTER 7

Markov Processes in Continuous Time

7.1 Introduction

So far we have been investigating the properties of stochastic processes in which both the basic random variable and the time parameter have been discrete. We now turn to processes which take place in continuous time instead of having distinct generations, as with the Markov chains of the last chapter. We shall continue to deal only with integral-valued random variables, though extensions to the diffusion type of process with continuous variables and continuous time will be made later, in Chapter 14. And we shall still retain the characteristic Markov restriction that, *given* the present state of the system, the future behavior is independent of the past history. Such processes will be called *Markov* processes.

Many problems involving applications to populations of individual units such as radioactive atoms, telephone calls, mutant genes, numbers of chromosome breakages, infectious persons, etc., all require the use of discrete random variables, though such problems are usually formulated so as to specify the size and behavior of the population at any instant of time. Only in rather special cases can we legitimately work with a fixed generation time and use the already discussed theory of Markov chains. It is therefore essential to develop methods of investigating processes which entail a continuous flow of probability distributions in time.

As a matter of general definition, we can suppose that the size of the population being considered is, at any instant of time t, represented by a discrete random variable $X(t)$, with

$$P\{X(t) = n\} = p_n(t), \quad n = 0, 1, 2, \cdots. \tag{7.1}$$

At certain instants of time there will be discrete changes in the population size, due for example to the loss or death of an individual, or to the appearance or birth of new individuals. To fix ideas let us consider a very simple type of process.

7.2 The Poisson process

Suppose we are measuring the radioactivity of a certain quantity of radioactive material using a Geiger counter. The counter actually records at any instant the total number of particles emitted in the interval from the time it was first set in operation. Let $X(t)$ be the total number of particles recorded up to time t, where this total number of particles is the population size in question. It is usual to assume that the events involved, i.e. the appearances of new radioactive particles, occur at random. Specifying what we mean by "random" quickly supplies a mathematical model of the whole process. Let us, for example, assume that the chance of a new particle being recorded in any short interval is independent, not only of the previous states of the system, but also of the present state (this would be a satisfactory assumption for periods of time that were relatively short compared with the half-life of the radioactive substance under investigation). We can therefore assume that the chance of a new addition to the total count during a very short interval of time Δt can be written as $\lambda \Delta t + o(\Delta t)$, where λ is some suitable constant characterizing the intensity of radioactivity. The chance of two or more simultaneous emissions will clearly be $o(\Delta t)$; and the chance of no change in the total count will be $1 - \lambda \Delta t - o(\Delta t)$. We can thus write

$$p_n(t + \Delta t) = p_{n-1}(t) \cdot \lambda \Delta t + p_n(t) \cdot (1 - \lambda \Delta t), \qquad (7.2)$$

(ignoring terms that are small compared with Δt), since if $n > 0$ the state involving precisely n events in the interval $(0, t + \Delta t)$, arises, *either* from $n - 1$ in the interval $(0, t)$ with one new emission in time Δt, *or* from n in the interval $(0, t)$ with no new emissions in Δt.

It follows immediately from (7.2) that

$$\frac{dp_n(t)}{dt} = \lim_{\Delta t \to 0} \frac{p_n(t + \Delta t) - p_n(t)}{\Delta t}$$

$$= \lambda \{ p_{n-1}(t) - p_n(t) \}, \qquad n > 0. \qquad (7.3)$$

The equation for $n = 0$ is obtained more simply, since $n = 0$ at time $t + \Delta t$ only if $n = 0$ at time t and no new particles are emitted in Δt. In this case we easily find

$$\frac{dp_0(t)}{dt} = -\lambda p_0(t). \qquad (7.4)$$

If we start off the process with the Geiger counter set to zero, the initial condition is

$$p_0(0) = 1. \tag{7.5}$$

The set of differential-difference equations (7.3) and (7.4), together with the boundary condition (7.5), are sufficient to determine the probability distribution $p_n(t)$. Thus we can solve (7.4) immediately to give, using (7.5),

$$p_0(t) = e^{-\lambda t}.$$

Substituting this in the first equation of (7.3) with $n = 1$ and solving gives

$$p_1(t) = \lambda t e^{-\lambda t}.$$

This procedure can now be repeated. Substituting the above value of $p_1(t)$ in the second equation of (7.3) with $n = 2$ gives a differential equation for $p_2(t)$. Solving in this way for successive $p_n(t)$ yields the general formula

$$p_n(t) = e^{-\lambda t}(\lambda t)^n/n!, \quad n = 0, 1, 2, \cdots. \tag{7.6}$$

This is of course simply a Poisson distribution with parameter λt, and gives rise to the designation of *Poisson process*. It is perhaps worth remarking in passing that in this derivation we obtain the Poisson distribution directly from the basic formulation of the mathematical model, and not indirectly as the limiting form of some other distribution such as the binomial. Moreover, we should obtain a Poisson distribution for any other situation involving random events, for which the basic model might still apply, e.g. accidents occurring in a specified period, or breaks developing in a given portion of a chromosome.

Next, we notice that $p_0(t) = e^{-\lambda t}$ is the chance that at time t no event has occurred, i.e. it is the chance that the *first* event happens at some instant greater than t. The distribution function of the arrival time u of the first event is accordingly given by

$$F(u) = 1 - e^{-\lambda u}, \tag{7.7}$$

and the corresponding frequency function is

$$f(u) \equiv F'(u) = \lambda e^{-\lambda u}, \quad 0 \leqslant u < \infty. \tag{7.8}$$

This negative exponential distribution is the distribution of the interval between any arbitrarily chosen point of time and the point at which the next event occurs; and in particular it is the distribution of the time interval between any two successive events.

The connection exhibited above between the negative exponential distribution of the interval between successive events, the Poisson distribution of the number of events in a fixed interval, and the "random" occurrence of the events, is of considerable practical importance, and we shall have occasion to make use of these properties in more complicated situations.

7.3 Use of generating functions

In the simplest possible set-up involving the Poisson process just discussed in the previous section the appropriate set of differential-difference equations given by (7.3) and (7.4) can be solved in a comparatively elementary manner. However, the technique of successive solution of the equations can lead to considerable difficulties when applied to more complicated processes. Appreciable advantages in handling the equations often result from a use of generating functions.

We define a probability-generating function $P(x, t)$ for the probability distribution $p_n(t)$ in the usual way as

$$P(x, t) = \sum_{n=0}^{\infty} p_n(t)x^n, \tag{7.9}$$

where $P(x, t)$ is a function of t as well as x.

To see how this approach can be applied to a treatment of the Poisson process, consider the basic differential-difference equations (7.3) and (7.4). Multiply the equation for n by x^n and sum over all values of n. This gives

$$\sum_{n=0}^{\infty} \frac{dp_n(t)}{dt} x^n = \lambda \sum_{n=1}^{\infty} p_{n-1}(t)x^n - \lambda \sum_{n=0}^{\infty} p_n(t)x^n,$$

or
$$\frac{\partial P(x, t)}{\partial t} = \lambda x P(x, t) - \lambda P(x, t)$$

$$= \lambda(x - 1)P(x, t), \tag{7.10}$$

with initial condition

$$P(x, 0) = 1. \tag{7.11}$$

Since (7.10) involves differentiation in t only, we can integrate directly to obtain, in conjunction with (7.11),

$$P(x, t) = e^{\lambda t(x-1)}. \tag{7.12}$$

This is immediately identifiable as the probability-generating function of a Poisson distribution with parameter λt, the individual probabilities being given by (7.6).

From (7.10) we can pass directly to an equation for the moment-generating function $M(\theta, t)$, merely by putting $x = e^{\theta}$, giving

$$\frac{\partial M(\theta, t)}{\partial t} = \lambda(e^{\theta} - 1)M(\theta, t). \qquad (7.13)$$

Similarly, writing $K(\theta, t) = \log M(\theta, t)$ for the cumulant-generating function, we have

$$\frac{\partial K(\theta, t)}{\partial t} = \lambda(e^{\theta} - 1). \qquad (7.14)$$

If desired we could easily solve (7.13) or (7.14), though in the present simple example the functions $M(\theta, t)$ and $K(\theta, t)$ can be derived directly from (7.12). In more advanced applications, however, one of the three partial differential equations, for the probability-generating function, the moment-generating function, and the cumulant-generating function, may be more easily soluble than the others.

Extensions and modifications of the above approach will be widely used in the sequel. It is, therefore, worth examining a number of devices that are specially useful in this connection.

7.4 "Random-variable" technique

One valuable procedure is the technique for writing down, more or less at sight, the appropriate partial differential equation for the probability-generating function or the moment-generating function. This method is based on a "random-variable" approach to the problem. Before deriving a fairly general result, let us first see how the method works in the special case of the Poisson process discussed in Section 7.2.

As before, we represent the size of the population of individuals under consideration at time t by the random variable $X(t)$. The corresponding variable at time $t + \Delta t$ is $X(t + \Delta t)$. Let the increment (which in general may be positive, negative, or zero) in $X(t)$ during time Δt be defined by

$$\Delta X(t) = X(t + \Delta t) - X(t). \qquad (7.15)$$

As in (7.9) we write the probability-generating function for $X(t)$ as $P(x, t)$, and for $X(t + \Delta t)$ as $P(x, t + \Delta t)$. Let us also write $\Delta P(x, t)$ for probability-generating function of $\Delta X(t)$.

Now in the case of the Poisson process the variables $X(t)$ and $\Delta X(t)$ are independently distributed. This follows immediately from the stipulation that the chance of appearance of a new particle at any instant is to be independent, not only of previous states of the system, but also of the present state. If the variables are independent, the probability-generating function of the sum is the product of the individual generating functions, i.e.

$$P(x, t + \Delta t) = P(x, t) \, \Delta P(x, t). \tag{7.16}$$

But from the first paragraph of Section 7.2 we obtain

$$\Delta P(x, t) = E\{x^{\Delta X(t)}\}$$
$$= (1 - \lambda \, \Delta t)x^0 + \lambda \, \Delta t x^1$$
$$= 1 + \lambda(x - 1) \, \Delta t, \tag{7.17}$$

to first order in Δt. Substitution of (7.17) in (7.16) now yields

$$P(x, t + \Delta t) = P(x, t)\{1 + \lambda(x - 1) \, \Delta t\}.$$

Hence

$$\frac{\partial P(x, t)}{\partial t} = \lim_{\Delta t \to 0} \frac{P(x, t + \Delta t) - P(x, t)}{\Delta t}$$
$$= \lambda(x - 1)P(x, t), \tag{7.18}$$

which simply recovers equation (7.10).

A similar argument for the moment-generating function, is, in an obvious notation, as follows. First, we can write

$$M(\theta, t + \Delta t) = M(\theta, t) \, \Delta M(\theta, t). \tag{7.19}$$

Secondly, we have

$$\Delta M(\theta, t) = E\{e^{\theta \Delta X(t)}\}$$
$$= (1 - \lambda \, \Delta t)e^0 + \lambda \, \Delta t e^{\theta}$$
$$= 1 + \lambda(e^{\theta} - 1) \, \Delta t. \tag{7.20}$$

Equations (7.19) and (7.20) can now be combined to yield (7.13).

The above derivations have been simplified by the fact that, for the Poisson process, the variables $X(t)$ and $\Delta X(t)$ are independent. This is, of course, not true in general, and so we cannot as a rule multiply the generating functions for $X(t)$ and $\Delta X(t)$. However, an extension of the above approach will produce a more general result, of which we shall make frequent use in subsequent discussions.

Let us suppose that a finite number of transitions are possible in the interval Δt. In fact let

$$P\{\Delta X(t) = j | X(t)\} = f_j(X)\Delta t, \quad j \neq 0, \tag{7.21}$$

where the $f_j(X)$ are suitable non-negative functions of $X(t)$, and j may be positive or negative (in some applications we could interpret positive values as births and negative values as deaths). The chance of no transitions is accordingly

$$P\{\Delta X(t) = 0 | X(t)\} = 1 - \sum_{j \neq 0} f_j(X)\,\Delta t. \tag{7.22}$$

Next, we distinguish more precisely between different operators for expectations. By $\underset{t}{E}$ and $\underset{t+\Delta t}{E}$ we mean the taking of the expectations of some function at time t with respect to variations in $X(t)$, and at time $t + \Delta t$ for variations in $X(t + \Delta t)$. And by $\underset{\Delta t|t}{E}$ we mean the calculation of a conditional expectation at the end of the interval Δt for variations in the variable $\Delta X(t)$, *given* the value of $X(t)$ at time t. If we use the usual formula for the definition of an expectation, it is easy to show that, for an arbitrary function $\Phi\{X(t)\}$, we may write

$$\underset{t+\Delta t}{E}\,[\Phi\{X(t + \Delta t)\}] = \underset{t}{E}\left\{\underset{\Delta t|t}{E}\,[\Phi\{X(t) + \Delta X(t)\}]\right\}. \tag{7.23}$$

This simply means that we can evaluate the expectation at $t + \Delta t$ in two successive stages, first considering the value for Δt given t, and then finding the expectation of the latter result at t.

It is slightly simpler to work with moment-generating functions than with probability-generating functions. Writing $\Phi = M$, and using (7.23) gives

$$M(\theta, t + \Delta t) = \underset{t+\Delta t}{E}\,\{e^{\theta X(t + \Delta t)}\}$$

$$= \underset{t+\Delta t}{E}\,\{e^{\theta X(t) + \theta\,\Delta X(t)}\}$$

$$= \underset{t}{E}\left[e^{\theta X(t)}\,\underset{\Delta t|t}{E}\,\{e^{\theta\,\Delta X(t)}\}\right]. \tag{7.24}$$

From (7.24) it follows that

$$\frac{\partial M}{\partial t} = \lim_{\Delta t \to 0}\frac{M(\theta, t + \Delta t) - M(\theta, t)}{\Delta t}$$

$$= \lim_{\Delta t \to 0}\frac{1}{\Delta t}\left[\underset{t}{E}\left\{e^{\theta X(t)}\,\underset{\Delta t|t}{E}\,\{e^{\theta\,\Delta X(t)}\}\right\} - \underset{t}{E}\,\{e^{\theta X(t)}\}\right]$$

$$= \underset{t}{E}\left[e^{\theta X(t)}\lim_{\Delta t \to 0}\underset{\Delta t|t}{E}\left\{\frac{(e^{\theta\,\Delta X(t)} - 1)}{\Delta t}\right\}\right]. \tag{7.25}$$

If therefore the conditional expectation, given $X(t)$ at t, of

$$\frac{e^{\theta \, \Delta X(t)} - 1}{\Delta t}$$

has a limit, say $\Psi(\theta, t, X)$, as $\Delta t \to 0$, we can express (7.25) as

$$\frac{\partial M(\theta, t)}{\partial t} = E_t \left\{ e^{\theta X(t)} \Psi(\theta, t, X) \right\}$$

$$= \Psi \left(\theta, t, \frac{\partial}{\partial \theta} \right) M(\theta, t), \qquad (7.26)$$

where the operator $\partial/\partial\theta$ acts only on $M(\theta, t)$, provided that the differential and expectation operators are commutable.

In the special case in which the conditional transition probabilities are given by (7.21) and (7.22), the function Ψ has the form

$$\Psi(\theta, t, X) = \lim_{\Delta t \to 0} E_{\Delta t | t} \left\{ \frac{e^{\theta \, \Delta X(t)} - 1}{\Delta t} \right\}$$

$$= \lim_{\Delta t \to 0} \frac{\left\{ 1 - \sum_{j \neq 0} f_j(X) \, \Delta t \right\} + \sum_{j \neq 0} f_j(X) \, \Delta t e^{j\theta} - 1}{\Delta t}$$

$$= \sum_{j \neq 0} (e^{j\theta} - 1) f_j(X). \qquad (7.27)$$

We can now write (7.26) as

$$\frac{\partial M(\theta, t)}{\partial t} = \sum_{j \neq 0} (e^{j\theta} - 1) f_j \left(\frac{\partial}{\partial \theta} \right) M(\theta, t). \qquad (7.28)$$

Putting $e^\theta = x$ and $\partial/\partial\theta = x\partial/\partial x$ in (7.28) gives the corresponding expression for the probability-generating function as

$$\frac{\partial P(x, t)}{\partial t} = \sum_{j \neq 0} (x^j - 1) f_j \left(x \frac{\partial}{\partial x} \right) P(x, t). \qquad (7.29)$$

It often happens that j takes only such values as $+1$ or -1, and that f_j is also simple in form. The formulas (7.28) and (7.29) then become very easy to apply. In the case of the Poisson process, for example, we have only $j = +1$, and $f_1(X) = \lambda$. Equation (7.10) can then be written down immediately. More elaborate applications will be made in later chapters.

It is easy to extend these results to situations involving two or more variables. Suppose, for instance, there are two random variables $X(t)$ and $Y(t)$, with joint transition probabilities given by

$$P\{\Delta X(t) = j, \Delta Y(t) = k|X(t), Y(t)\} = f_{jk}(X, Y)\,\Delta t, \qquad (7.30)$$

excluding the case of *both* j and k being zero together. The result corresponding to (7.28) is now

$$\frac{\partial M(\theta, \phi, t)}{\partial t} = \sum (e^{j\theta + k\phi} - 1)f_{jk}\left(\frac{\partial}{\partial \theta}, \frac{\partial}{\partial \phi}\right) M(\theta, \phi, t), \quad j, k \text{ not both zero.}$$

$$(7.31)$$

(See Chapter 10 for applications of this last result.)

7.5 Solution of linear partial differential equations

For many processes the functions f_j or f_{jk} in (7.28) and (7.31) are merely linear functions of the differential operators $\partial/\partial\theta$ or $\partial/\partial\theta$ and $\partial/\partial\phi$, respectively. In such cases the basic partial differential equation is linear, and an explicit solution is often not too difficult to obtain. Difficulties arise as soon as f_j or f_{jk} is a quadratic function, since this leads to a partial differential equation of second order, which is liable to be somewhat intractable in many cases of special interest, e.g. epidemics. For an account of the general theory of such equations the reader should consult some standard textbook such as Forsyth (1929, p. 392). But since we shall make frequent use in the sequel of models leading to simple *linear* partial differential equations, it is perhaps worth recalling here the main results that we shall require. (Certain more specialized equations will be dealt with ad hoc as they occur.)

Suppose we have the equation

$$P\frac{\partial z}{\partial x} + Q\frac{\partial z}{\partial y} = R, \qquad (7.32)$$

subject to some appropriate boundary conditions, where P, Q, and R may all be functions of x, y, and z. The first step is to form the subsidiary equations given by

$$\frac{dx}{P} = \frac{dy}{Q} = \frac{dz}{R}. \qquad (7.33)$$

We now find two independent integrals of these subsidiary equations, writing them in the form

$$u(x, y, z) = \text{constant}, \qquad v(x, y, z) = \text{constant}. \qquad (7.34)$$

The most general solution of (7.32) is now given by

$$\left.\begin{array}{c} \Phi(u, v) = 0 \\ u = \Psi(v) \end{array}\right\}, \qquad (7.35)$$

or

where Φ and Ψ are arbitrary functions.

Although it might seem that the appearance of the arbitrary functions Φ or Ψ involves a result too general to be of use to us, the precise form of these functions is determined, in all cases of interest, by an appeal to the boundary conditions.

More generally, suppose we have the linear equation

$$\sum_{j=1}^{n} P_j \frac{\partial z}{\partial x_j} = R, \qquad (7.36)$$

where the P_j and R are functions of $x_1, \cdots, x_n$, and z. The analog of (7.33) is

$$\frac{dx_1}{P_1} = \cdots = \frac{dx_n}{P_n} = \frac{dz}{R}. \qquad (7.37)$$

We must now find n independent integrals of (7.37), namely

$$u_j(x_1, \cdots, x_n, z) = \text{constant}, \qquad j = 1, 2, \cdots, n, \qquad (7.38)$$

and the most general solution takes the form

$$\Phi(u_1, \cdots, u_n) = 0. \qquad (7.39)$$

7.6 General theory

So far in this chapter we have approached the problem of a stochastic process in continuous time in a comparatively simple and direct, though somewhat intuitive, way. However, more insight is afforded into the structure of such processes by developing a more general theory. This we shall now do, but we shall avoid discussion of the finer points of rigor in order not to obscure the consequences for practical applications.

It is convenient to adopt a notation which is in many ways analogous to that already employed in the treatment of Markov chains in Chapter 5. Thus, where we previously wrote $p_{ij}^{(n)}$ for the n-step transition from state E_i to state E_j, we now write $p_{ij}(s, t)$ for the conditional probability of finding the system in E_j at time t, *given* that it was previously in E_i at time s, i.e.

$$P\{X(t) = j | X(s) = i\} = p_{ij}(s, t). \qquad (7.40)$$

Now consider three epochs of time s, t, and u, where $s < t < u$. Let the system be in state E_i at time s and E_n at time u. Then the course followed by an actual process must go via some state, say E_j, at time t. For a Markov type of process the probability of this particular path is $p_{ij}(s, t) p_{jn}(t, u)$, since the two component probabilities are by definition independent. The probability of E_n at u, given E_i at s, must therefore be given by summing over all values of j, i.e.

$$p_{in}(s, u) = \sum_j p_{ij}(s, t)p_{jn}(t, u). \qquad (7.41)$$

The equations given by (7.41) for different i and n are the *Chapman–Kolmogorov equations*.

As with Markov chains in discrete time, many of the results for Markov processes in continuous time can be expressed more succinctly by the use of a matrix notation. Let us write

$$\mathbf{P}(t|s) \equiv \{p_{ij}(s, t)\}', \qquad (7.42)$$

where, as in Section 5.2, we define the transition matrix as the transpose of $\{p_{ij}\}$. We are thus able to preserve the usual convention for the order of suffices in $\mathbf{P}$, and have the columns as conditional distributions. The Chapman–Kolmogorov equations in (7.41) now take the form

$$\mathbf{P}(u|s) = \mathbf{P}(u|t)\mathbf{P}(t|s). \qquad (7.43)$$

So far we have not made any assumptions about homogeneity in time, but if the process were homogeneous we could write

$$p_{ij}(s, t) = p_{ij}(t - s), \qquad (7.44)$$

where the transition probabilities would depend only on the time difference between the two epochs s and t. For two successive time intervals of length σ and τ, we could write (7.43), in an obvious but slightly different notation, as

$$\mathbf{P}(\sigma + \tau) = \mathbf{P}(\tau)\mathbf{P}(\sigma). \qquad (7.45)$$

This equation is clearly analogous to the rule for the multiplication of transition matrices in homogeneous Markov chains (see Section 5.2).

Next, let us write the absolute distribution at time s as the column vector $\{p_i(s)\} \equiv \mathbf{p}(s)$, and the corresponding distribution at time t as $\{p_j(t)\} \equiv \mathbf{p}(t)$. Then it is clear that in general we must have

$$p_n(t) = \sum_i p_i(s)p_{in}(s, t), \qquad (7.46)$$

or, in matrix notation,

$$\mathbf{p}(t) = \mathbf{P}(t|s)\mathbf{p}(s), \qquad (7.47)$$

since the probability that the system is in state E_i at time s is $p_i(s)$, and the chance of a transition from E_i at s to E_n at t is $p_{in}(s, t)$, and we must sum over all such transitions leading to E_n at t. Equation (7.47) is analogous to the result (5.9) for Markov chains.

We now undertake the derivation of certain fundamental differential equations for the process by considering the Chapman–Kolmogorov equation in (7.41), and keeping the first two epochs s and t fixed, while putting $u = t + \Delta t$, and then seeing what happens as $\Delta t \to 0$. We have immediately

$$p_{in}(s, t + \Delta t) = \sum_j p_{ij}(s, t)p_{jn}(t, t + \Delta t)$$

$$= p_{in}(s, t)p_{nn}(t, t + \Delta t) + \sum_{j \neq n} p_{ij}(s, t)p_{jn}(t, t + \Delta t). \quad (7.48)$$

Now if the system is in state E_n at time t, $p_{nn}(t, t + \Delta t)$ is the probability of no change in the interval Δt. Thus $1 - p_{nn}(t, t + \Delta t)$ is the probability of some change in the interval, and we may reasonably suppose that this quantity may be written as $\lambda_n(t) \Delta t$. Therefore

$$\lim_{\Delta t \to 0} \frac{1 - p_{nn}(t, t + \Delta t)}{\Delta t} = \lambda_n(t). \quad (7.49)$$

Next, we suppose that there exist limiting transition probabilities $q_{jn}(t)$, which are *conditional* on some change having occurred. That is, if the system is in state E_j at time t, and if a change occurs, then the chance of a transition to E_n is $q_{jn}(t)$. Thus for $j \neq n$ we have

$$p_{jn}(t, t + \Delta t) = \lambda_j(t)\Delta t \cdot q_{jn}(t),$$

from which it follows that

$$\lim_{\Delta t \to 0} \frac{p_{jn}(t, t + \Delta t)}{\Delta t} = \lambda_j(t)q_{jn}(t), \quad j \neq n. \quad (7.50)$$

We now obtain from (7.48), using both (7.49) and (7.50),

$$\frac{\partial p_{in}(s, t)}{\partial t} = \lim_{\Delta t \to 0} \frac{p_{in}(s, t + \Delta t) - p_{in}(s, t)}{\Delta t}$$

$$= -\lambda_n(t)p_{in}(s, t) + \sum_{j \neq n} p_{ij}(s, t)\lambda_j(t)q_{jn}(t). \quad (7.51)$$

These are Kolmogorov's system of *forward differential equations*, and they refer to changes in the final time t. It may be noticed that the system really consists of ordinary differential equations, so far as the search for a solution is concerned, since both i and s are regarded as fixed. The required initial conditions are obviously

$$p_{in}(s, s) = 1, \quad n = i \left.\vphantom{\begin{matrix}a\\b\end{matrix}}\right\}.$$
$$\phantom{p_{in}(s, s)} = 0, \quad n \neq i \left.\vphantom{\begin{matrix}a\\b\end{matrix}}\right\} \tag{7.52}$$

In an entirely analogous manner we can investigate the limit

$$\lim_{\Delta t \to 0} \frac{p_{in}(s - \Delta s, t) - p_{in}(s, t)}{\Delta s}$$

and obtain the *backward differential equations*

$$\frac{\partial p_{in}(s, t)}{\partial s} = \lambda_i(s) p_{in}(s, t) - \lambda_i(s) \sum_{j \neq i} q_{ij}(s) p_{jn}(s, t), \tag{7.53}$$

with initial conditions

$$p_{in}(t, t) = 1, \quad i = n \left.\vphantom{\begin{matrix}a\\b\end{matrix}}\right\}.$$
$$\phantom{p_{in}(t, t)} = 0, \quad i \neq n \left.\vphantom{\begin{matrix}a\\b\end{matrix}}\right\} \tag{7.54}$$

The Kolmogorov backward and forward equations can be written, and indeed derived, more compactly by a matrix method. Thus suppose that the probability of some change in the interval Δt out of state E_n is specified by the nth element in the diagonal matrix $\mathbf{\Lambda}(t) \, \Delta t$, and let the limiting conditional transition matrix be $\mathbf{Q}(t)$. Then from (7.43) we have

$$\mathbf{P}(t + \Delta t | s) = \mathbf{P}(t + \Delta t | t) \mathbf{P}(t | s)$$

$$= \{\mathbf{I} - \mathbf{\Lambda}(t) \, \Delta t\} \mathbf{P}(t | s) + \{\mathbf{Q}(t) \mathbf{\Lambda}(t) \, \Delta t\} \mathbf{P}(t | s).$$

Hence the forward equation is

$$\frac{\partial \mathbf{P}(t | s)}{\partial t} = \{\mathbf{Q}(t) - \mathbf{I}\} \mathbf{\Lambda}(t) \mathbf{P}(t | s)$$

$$= \mathbf{R}(t) \mathbf{P}(t | s), \tag{7.55}$$

where $$\mathbf{R}(t) = \{\mathbf{Q}(t) - \mathbf{I}\} \mathbf{\Lambda}(t). \tag{7.56}$$

By a similar argument the backward equation comes out as

$$\frac{\partial \mathbf{P}(t | s)}{\partial s} = -\mathbf{P}(t | s) \mathbf{R}(s). \tag{7.57}$$

The correspondence between (7.55) and (7.51); and between (7.57) and (7.53); means that if $\mathbf{R}(t) \equiv \{r_{ij}(t)\}$, then $r_{ii}(t) = -\lambda_i(t)$, and $r_{ij}(t) = q_{ji}(t) \lambda_j(t)$ $(i \neq j)$. The quantities $r_{ij} \, \Delta t$ are called the *infinitesimal transition probabilities*.

If we now postmultiply each side of the forward equation (7.55) by the vector $\mathbf{p}(s)$ and use (7.47), we obtain

$$\frac{\partial \mathbf{p}(t)}{\partial t} = \mathbf{R}(t)\mathbf{p}(t), \tag{7.58}$$

which provides a basic set of differential-difference equations for the absolute probabilities $\mathbf{p}(t)$ at time t.

In the homogeneous case where the $p_{ij}(s, t)$ depend only on the difference $t - s$ (see equation (7.44)), it is clear that the λ_n and q_{jn} are constants independent of t. The forward equation (7.51) can then be written

$$\frac{dp_{in}(t)}{dt} = -\lambda_n p_{in}(t) + \sum_{j \neq n} \lambda_j q_{jn} p_{ij}(t), \tag{7.59}$$

and a similar argument applied to the backward equation (7.53) leads to the form

$$\frac{dp_{in}(t)}{dt} = -\lambda_i p_{in}(t) + \lambda_i \sum_{j \neq n} q_{ij} p_{jn}(t), \tag{7.60}$$

where it should be noted that we are using the differential coefficient with respect to t (and not s). In matrix form the two equations (7.59) and (7.60), give the forward expression

$$\frac{d\mathbf{P}(t)}{dt} = \mathbf{R}\mathbf{P}(t), \tag{7.61}$$

and the backward expression

$$\frac{d\mathbf{P}(t)}{dt} = \mathbf{P}(t)\mathbf{R}, \tag{7.62}$$

which also follow as the homogeneous versions of (7.55) and (7.57). We now have $\mathbf{R} = \{r_{ij}\}$, with $r_{ii} = -\lambda_i$ and $r_{ij} = q_{ji}\lambda_j$ $(i \neq j)$, independently of the value of t. The initial conditions for both of the last two equations are of course $\mathbf{P}(0) = \mathbf{I}$.

Finally, the absolute probabilities $\mathbf{p}(t)$ at time t must satisfy

$$\frac{d\mathbf{p}(t)}{dt} = \mathbf{R}\mathbf{p}(t), \tag{7.63}$$

which can be obtained directly from (7.58) for the special case of a homogeneous process.

A formal solution

$$\mathbf{P}(t) = e^{\mathbf{R}t} \tag{7.64}$$

for equations (7.61) and (7.62) suggests itself, together with the absolute distribution

$$\mathbf{p}(t) = e^{\mathbf{R}t}\mathbf{p}(0). \tag{7.65}$$

If there is only a finite number of states, so that $\mathbf{R}$ is a finite matrix, the exponential expansion in powers of $\mathbf{R}$ will be valid and we could make use of a spectral resolution technique (see Section 5.5). Thus if

$$\mathbf{R} = \sum_{i=1}^{k} \alpha_i \mathbf{A}_i,$$

where the $\mathbf{A}_i$ are a spectral set, then

$$\exp(\mathbf{R}t) = \sum_{i=1}^{k} e^{\alpha_i t} \mathbf{A}_i.$$

Now it can be shown that in general there always exists a common solution of the Kolmogorov forward and backward differential equations satisfying the initial conditions, and also the Chapman–Kolmogorov equation. It may happen, however, that the solution is not a true probability distribution in the sense that $\sum_j p_{ij}(s, t) < 1$. When this occurs we interpret $1 - \sum_j p_{ij}(s, t)$ as the probability of an infinity of transitions occurring in the interval (s, t). In practical applications this is hardly an aspect with which we need be concerned, though it could happen that we inadvertently chose a mathematical model with unrealistic properties such as "explosive" or "divergent" growth (see Section 8.4).

Just as with Markov chains in discrete time, so in the present case of continuous processes there are limiting properties to be considered. If, for a homogeneous process, the limits

$$\lim_{t \to \infty} p_{in}(t) = p_n \tag{7.66}$$

exist independent of i, and if $\{p_n\}$ is a probability distribution, then we say that the process is *ergodic* (compare Section 5.4 for Markov chains). Moreover, if the process is ergodic it can be shown that

$$\lim_{t \to \infty} p_n(t) = p_n \tag{7.67}$$

independently of the initial distribution $p_n(0)$. In this case the limiting distribution is uniquely determined by solving the linear system of equations

$$\mathbf{R}\mathbf{p} = 0, \tag{7.68}$$

obtained by putting $d\mathbf{p}/dt = 0$ in (7.63). This system of equations can be written explicitly as

$$\lambda_n p_n = \sum_{j \neq n} \lambda_j q_{jn} p_j, \tag{7.69}$$

using the definition of the elements of $\mathbf{R}$ following equation (7.62).

A special theorem for the case of a finite number of states is that if every state can be reached from every other state with positive probability, then the process is ergodic with limiting distribution given by the solution of (7.68) or (7.69).

Let us now return to a discussion of the Poisson process, which we dealt with earlier in this chapter in a rather intuitive way, and see how some of the general theory can be applied. The chance of an event occurring in time Δt is $\lambda \Delta t$. Thus the diagonal matrix $\Lambda(t)$, specifying the chance of some change in each state during Δt, is

$$\Lambda(t) = \begin{bmatrix} \lambda & & 0 & \\ & \lambda & & \\ 0 & & \lambda & \\ & & & \ddots \end{bmatrix}, \qquad (7.70)$$

where all non-diagonal elements are zero. This precise form arises because the chance $\lambda \Delta t$ applies to every state independently of the state. We next consider the form of the matrix of conditional transition probabilities $\mathbf{Q}(t)$. *Given* that a transition has occurred, we know that it can involve only the addition of a simple unit to the total count, i.e. it must be of type $E_n \rightarrow E_{n+1}$. We accordingly have

$$\mathbf{Q}(t) = \begin{bmatrix} 0 & & & 0 & \\ 1 & 0 & & & \\ & 1 & 0 & & \\ & & 1 & 0 & \\ 0 & & & 1 & \ddots \end{bmatrix}, \qquad (7.71)$$

where the diagonal immediately below the leading diagonal consists entirely of unit elements, and all other elements are zero. Thus $\mathbf{R} = (\mathbf{Q} - \mathbf{I})\Lambda$ is given by

$$\mathbf{R}(t) = \lambda \begin{bmatrix} -1 & & & 0 & \\ 1 & -1 & & & \\ & 1 & -1 & & \\ & & 1 & & \ddots \\ 0 & & & & \end{bmatrix}. \qquad (7.72)$$

Equation (7.63) for the absolute state probabilities is therefore equivalent to the set of equations

$$\left. \begin{aligned} \frac{dp_0}{dt} &= -\lambda p_0 \\[2mm] \frac{dp_n}{dt} &= \lambda p_{n-1} - \lambda p_n, \quad n > 0 \end{aligned} \right\} \qquad (7.73)$$

as already given in (7.3). The Kolmogorov forward and backward equations are easily found to be

$$\frac{dp_{ij}(t)}{dt} = -\lambda p_{ij}(t) + \lambda p_{i,j-1}(t), \qquad (7.74)$$

and
$$\frac{dp_{ij}(t)}{dt} = -\lambda p_{ij}(t) + \lambda p_{i+1,j}(t). \qquad (7.75)$$

It can be verified quite simply that the appropriate solution of these equations is, subject to the initial condition $p_{ij}(0) = \delta_{ij}$,

$$\left. \begin{aligned} p_{ij}(t) &= \frac{e^{-\lambda t}(\lambda t)^{j-1}}{(j-1)!}, \quad j \geqslant i \\ &= 0, \qquad\qquad\quad j < i \end{aligned} \right\}, \qquad (7.76)$$

as we should expect from the general properties of the Poisson process already obtained.

In many practical applications it is the forward equation which is obtained most naturally, and is therefore the one most frequently used for finding the required probability distributions. However, the backward equation can also be extremely valuable in certain contexts, and its possible importance in any problem should not be lost sight of (see for example Section 8.9).

PROBLEMS FOR SOLUTION

1. Suppose that the disintegration of radioactive atoms in a certain quantity of radioactive material follows a Poisson process with parameter λ as described in Section 7.2. Suppose, in addition, that there is a probability p of each particle emitted actually being recorded, the probabilities being independent.

Show that the events actually recorded follow a Poisson process with parameter λp.

2. If the random variables $X(t)$ and $Y(t)$ represent two independent Poisson processes with parameters λ and μ, prove that the process defined by

$$S(t) = X(t) + Y(t)$$

is also Poisson in character with parameter $\lambda + \mu$.

3. Let $X(t)$ be a random variable representing a Poisson process with parameter λ.

Derive the correlation coefficient of $X(t)$ and $X(t + \tau)$.

4. If the time interval between any two successive events in a certain stochastic process has a negative exponential distribution, independently of any other time interval, show that the process is Poisson.

5. Suppose that $X(t)$ represents the total number of individuals in a population at time t. The chance that any individual gives birth to another in time Δt is $\lambda \Delta t$. There is no death and the individuals reproduce independently. Let $X(0) = a$. Obtain a set of differential-difference equations for the probabilities $p_n(t) = P\{X(t) = n\}$. Solve these to obtain the quantities $p_n(t)$ explicitly.

Use the technique of Section 7.4 to write down a partial differential equation for the probability-generating function

$$P(x, t) = \sum_{n=0}^{\infty} p_n(t)x^n,$$

and solve this equation as indicated in Section 7.5.

[See also discussion of Section 8.2 later.]

6. Consider the Kolmogorov forward and backward equations given in (7.74) and (7.75). Solve these equations directly to obtain the solution given in (7.76).

7. Events occur independently in a certain stochastic process. The chance of an event occurring in the time interval $(t, t + \Delta t)$ is independent of the behavior of the process prior to time t and equals $ae^{-t/b}\Delta t + o(\Delta t)$. Prove that the chance of no event occurring in the interval (u, v) is $\exp\{-ab(e^{-u/b} - e^{-v/b})\}$.

CHAPTER 8

Homogeneous Birth and Death Processes

8.1 Introduction

In Chapter 7 we saw, first, how a fairly intuitive approach to the problem of Markov processes in continuous time could be used to develop some simple partial differential equations for the state probabilities, or, more concisely, for the probability- or moment-generating functions. We then saw how a more general theory could be elaborated, but examined in detail only one particular process, namely, the Poisson process for which new events occur entirely at random and quite independently of the past or even the present state of the system. However, in a great many processes of practical importance the appearance of new individuals, i.e. birth; or the disappearance of existing individuals, i.e. death; does depend in some degree at least on the present population size. We shall now study several situations of this type. In order to avoid mathematical intractability certain simplifying assumptions have to be made. Nevertheless, some insight is gained into the stochastic behavior of populations, and the simpler models can be used as a basis for developing more realistic extensions later. In particular, we shall confine the discussions in the present chapter to processes which are time-homogeneous.

8.2 The simple birth process

Let us suppose that we have a population of individuals whose total number at time t is given by the discrete random variable $X(t)$, where the probability that $X(t)$ takes the value n is $p_n(t)$. Further, let us assume that all individuals are capable of giving birth to new individuals. At the moment of birth we might suppose, either that the parent organism continues to exist along with an entirely new individual (as in the higher

84

animals), or that two new daughter organisms come into being, the reproducing individual ceasing to exist as such (as occurs with many unicellular organisms). In either case the total population size increases by exactly one unit. We now make the considerable, but convenient, assumption that the chance of a *given* individual producing a new one in time Δt is $\lambda \Delta t$. In the case of a single-celled organism, like a bacterium, this is equivalent to assuming that the time u elapsing between birth and the division into two daughter cells has the negative exponential distribution

$$f(u) = \lambda e^{-\lambda u}, \quad 0 \leqslant u < \infty. \tag{8.1}$$

(This follows from the arguments of Section 7.2, which apply equally well to the present case so long as we are considering the history of a single individual from birth to reproduction.) The adoption of a negative exponential distribution of life-times is a simplification which needs to be carefully examined in any practical application. Many bacteria, for example, reproduce neither randomly with negative exponential life-times, nor deterministically with fixed generation-times, but achieve an intermediate status with moderately variable life-times (see Section 10.5 for further discussion of these matters). Nevertheless, we shall investigate the present type of birth process with some interest to determine the consequences of the assumptions made.

Now if the chance of any given individual reproducing in time Δt is $\lambda \Delta t$, the corresponding probability that the whole population of size $X(t)$ will produce a birth is $\lambda X(t) \Delta t$ to first order in Δt. Following the approach adopted at the beginning of Section 7.2, we can readily write the probability of a population size n at time $t + \Delta t$ as

$$p_n(t + \Delta t) = p_{n-1}(t)\lambda(n - 1)\Delta t + p_n(t)(1 - \lambda n \Delta t),$$

leading to the differential-difference equation

$$\frac{dp_n}{dt} = \lambda(n - 1)p_{n-1} - \lambda n p_n. \tag{8.2}$$

If the population were zero, no births could occur, since new individuals can arise only from previously existing individuals, and the probability $\lambda X(t) \Delta t$ is zero if $X(t)$ vanishes. Thus we are concerned only with processes which start with a non-zero population size, say $X(0) = a$. The initial condition is then

$$p_a(0) = 1, \tag{8.3}$$

and the equations in (8.2) are required only for $n \geqslant a$ with $p_{a-1} \equiv 0$ in the first equation.

Now•the equations (8.2) *can* be solved in succession starting with the first one $dp_a/dt = -\lambda ap_a$. It is, however, considerably simpler to use the method of generating functions as suggested in Section 7.3. The required partial differential equation can be derived fairly easily from (8.2). An even more economical procedure is to apply the random-variable technique of Section 7.4. In the terminology of that section, there is only one kind of transition to be considered, i.e. $j = 1$, and the function $f_{+1}(X)$ is simply λX. Hence the partial differential equation for the moment-generating function can be written down immediately from (7.28) as

$$\frac{\partial M}{\partial t} = \lambda(e^\theta - 1)\frac{\partial M}{\partial \theta}, \tag{8.4}$$

the initial condition being, in conformity with (8.3),

$$M(\theta, 0) = e^{a\theta}. \tag{8.5}$$

We now have a partial differential equation of the simple linear type already discussed in Section 7.5. In fact (8.4) is precisely the same as (7.32) with the substitutions

$$x = t, \qquad y = \theta, \qquad z = M, \qquad P = 1, \qquad Q = -\lambda(e^\theta - 1), \qquad R = 0. \tag{8.6}$$

The subsidiary equations in (7.33) accordingly take the form

$$\frac{dt}{1} = \frac{d\theta}{-\lambda(e^\theta - 1)} = \frac{dM}{0}. \tag{8.7}$$

The first and third items of (8.7) give the integral

$$M = \text{constant}, \tag{8.8}$$

while the first and second give the equation

$$dt = \frac{e^{-\theta}\,d\theta}{-\lambda(1 - e^{-\theta})}.$$

This integrates immediately to yield

$$t = -\frac{1}{\lambda}\log(1 - e^{-\theta}) + \text{constant},$$

or $$e^{\lambda t}(1 - e^{-\theta}) = \text{constant}, \tag{8.9}$$

which is now in the form required for (7.34).

The general solution of (8.4) can therefore be written as

$$M = \Psi\{e^{\lambda t}(1 - e^{-\theta})\}, \tag{8.10}$$

using the two independent integrals in (8.8) and (8.9), where Ψ is an arbitrary function to be determined by the initial condition (8.5). Thus putting $t = 0$ in (8.10) we must have

$$e^{a\theta} = M(\theta, 0) = \Psi(1 - e^{-\theta}).\tag{8.11}$$

Let us now write $u = 1 - e^{-\theta}$, i.e. $e^{\theta} = (1 - u)^{-1}$. Substituting in (8.11) gives the desired form

$$\Psi(u) = (1 - u)^{-a}.\tag{8.12}$$

Applying (8.12) to (8.10) finally yields the solution

$$M(\theta, t) = \{1 - e^{\lambda t}(1 - e^{-\theta})\}^{-a}.\tag{8.13}$$

The probability-generating function is therefore

$$P(x, t) = \{1 - e^{\lambda t}(1 - x^{-1})\}^{-a}.\tag{8.14}$$

Thus the population size at time t has a negative binomial distribution (this is not the only way, however, in which such a distribution can arise—see Sections 8.7 and 9.2). Picking out the coefficient of x^n on the right of (8.14) gives

$$p_n(t) = \binom{n-1}{a-1} e^{-a\lambda t}(1 - e^{-\lambda t})^{n-a}, \quad n \geqslant a.\tag{8.15}$$

The stochastic mean and variance for this distribution are of course given by

$$m(t) = ae^{\lambda t},\tag{8.16}$$

and

$$\sigma^2(t) = ae^{\lambda t}(e^{\lambda t} - 1).\tag{8.17}$$

It may be noted that the mean value is precisely equal to the value obtained for a deterministically growing population of size $n(t)$ at time t, in which the increment in time Δt is $\lambda n \, \Delta t$.

The variance also increases steadily, and for large t is approximately $ae^{2\lambda t}$.

The above stochastic birth process was first studied by Yule (1924) in connection with the mathematical theory of evolution. The individuals of the population were thought of as the species within a genus, and the creation of a new species by mutation was conceived as being a random event with probability proportional to the number of species. Such a model neglects the differences in species sizes, and ignores the possibility of a species becoming extinct. The pure birth process can therefore be only a very rough first approximation.

A similar model was also used by Furry (1937) in dealing with cosmic ray phenomena, but again the model is rather inexact.

8.3   The general birth process

The foregoing discussion of the simple Yule–Furry type of birth process can be generalized by assuming that the chance of a birth in the interval Δt is $\lambda_n \Delta t$, where n is the population size at time t. The simple birth process of Section 8.2 is the special case in which the function λ_n is given by $\lambda_n = \lambda n$. In the present more general situation we easily find in the usual way that the basic differential-difference equations are

$$\left.\begin{aligned} \frac{dp_n}{dt} &= \lambda_{n-1}p_{n-1} - \lambda_n p_n, \quad n \geqslant 1 \\[2mm] \frac{dp_0}{dt} &= -\lambda_0 p_0 \end{aligned}\right\}. \tag{8.18}$$

Equations (8.2) are thus the particular form taken by (8.18) when $\lambda_n = \lambda n$.

We cannot present the solution of (8.18) in quite such a compact form as that for (8.2) owing to the more general form of the population birth-rates λ_n. However, we can solve the equations successively to obtain explicit expressions for the $p_n(t)$. Alternatively, we could use the Laplace transform approach which gives a somewhat neater derivation. See, for example, Bartlett (1955, Section 3.2).

Such a process could be used to describe the growth of a population in which, either λ_n was some specified algebraic function of the population size n, or the numerical values of the λ_n were specified for all population sizes.

More generally, we could employ the actual values taken by the random variable $X(t)$ to refer to states of the system which were not necessarily population sizes. Thus in applications to the theory of radioactive transmutations, we might use E_0 to refer to the initial radioactive uranium atom. This atom would then change successively, by the emission of α-particles or γ-rays, through a series of other atoms represented by the chain of states

$$E_0 \rightarrow E_1 \rightarrow E_2 \rightarrow \cdots \rightarrow E_r, \tag{8.19}$$

where E_r is the terminal state. The probability of a transmutation from E_n to E_{n+1} in the interval Δt would be $\lambda_n \Delta t$, while the terminal probability λ_r would of course be zero. The state E_r is an absorbing state, and we can take $p_n(t) \equiv 0$, $n > r$. In such an application there is no question of $X(t)$ growing indefinitely, the process coming to a halt as soon as the absorbing state is reached.

8.4 Divergent birth processes

As we saw in Section 8.2, one of the chief features of the simple birth process is that the stochastic mean increases steadily at an exponential rate, just like the corresponding value for a purely deterministic growth. Although we often find that stochastic means are equal to the exact values in the deterministic analog, this is by no means always the case (see, for example, queues and epidemics).

Consider for a moment a deterministic process in which the birth-rate *per individual* is proportional to the population size. Thus treating the population as a continuous variable, each unit would increase by an amount $\lambda n \, \Delta t$ in time Δt. And the whole population of size n would be subject to an increment of $\lambda n^2 \Delta t$ in Δt. The rate of growth would therefore be given by the differential equation

$$\frac{dn}{dt} = \lambda n^2, \tag{8.20}$$

for which the solution, subject to the initial condition $n(0) = a$, is

$$n(t) = \frac{a}{1 - \lambda a t}. \tag{8.21}$$

It can be seen from (8.21) that

$$\lim_{t \to 1/(\lambda a)} n(t) = \infty. \tag{8.22}$$

This means that a population growth-rate equal to the square of the population size entails an infinite growth in a finite time. We can call this type of growth *divergent* or *explosive*. In most biological situations this type of model would be inappropriate, though in certain circumstances where a very rapid rate of growth was involved it might be used as a first approximation.

We are, of course, primarily interested here in the properties of stochastic models. And by analogy with the deterministic set-up discussed above we might expect a stochastic birth process to be divergent in some sense if the birth-rates λ_n were not suitably restricted. We have already seen in the general theoretical discussion of Section 7.6 that the solution of the Kolmogorov equations may produce transition probabilities $p_{ij}(s, t)$ for which

$$\sum_j p_{ij}(s, t) < 1,$$

and that this contingency, when it occurs, is interpretable as meaning that there is a non-zero probability $1 - \sum_j p_{ij}(s, t)$ of an infinite number of transitions occurring from state E_i in the finite interval of time $t - s$. This we may regard as defining divergent growth in the stochastic case.

Similarly, if for the type of birth process investigated above the absolute distribution of probabilities at time t is such that

$$\sum_n p_n(t) < 1,$$

we shall have a divergent or explosive mode of growth. It is not difficult to show that the requirement for non-explosiveness, i.e.

$$\sum_n p_n(t) = 1, \quad \text{for all } t, \tag{8.23}$$

has, as a necessary and sufficient condition the divergence of

$$\sum_n \lambda_n^{-1}. \tag{8.24}$$

For an easy proof of this result see Feller (1957, Chapter 17, Section 4). We thus have a rule for deciding whether a particular model, which seems reasonable in other respects, in fact has the probably undesirable property of explosiveness.

8.5 The simple death process

The converse of the simple birth process investigated in Section 8.2 is the simple death process. In this we envisage a population of individuals which is subject to random death, in the sense that the probability of any individual dying in time Δt is $\mu \, \Delta t$. With such an assumption we have of course the corollary that individual life-times u have the negative exponential distribution given by

$$f(u) = \mu e^{-\mu u}, \quad 0 \leqslant u < \infty. \tag{8.25}$$

If we again use the random variable $X(t)$ to represent the total population size at time t, then the chance of one death occurring in the whole population in the interval Δt is $\mu X(t) \, \Delta t$. In the terminology of Section 7.4, there is just one type of transition with $j = -1$, and $f_{-1}(X)$ is then μX. The appropriate partial differential equation for the moment-generating function of the process is therefore

$$\frac{\partial M}{\partial t} = \mu(e^{-\theta} - 1)\frac{\partial M}{\partial \theta}, \tag{8.26}$$

with initial condition

$$M(\theta, 0) = e^{a\theta}, \tag{8.27}$$

if we start with a individuals at $t = 0$.

Comparison of (8.26) and (8.27) with (8.4) and (8.5) shows that the equations specifying the simple birth process can be transformed into those for the simple death process by substituting $-\mu$, $-\theta$, and $-a$ for λ, θ, and a, respectively. (Note that this entails replacing $\partial/\partial\theta$ by $-\partial/\partial\theta$ as well.) The solution given in (8.13) can therefore be immediately transformed into the required solution of (8.26). It is

$$M(\theta, t) = \{1 - e^{-\mu t}(1 - e^{\theta})\}^{a}. \tag{8.28}$$

The probability-generating function is thus

$$P(x, t) = \{1 - e^{-\mu t}(1 - x)\}^{a}, \tag{8.29}$$

so that the population size at time t has an ordinary binomial distribution, the individual probabilities being given by

$$p_{n}(t) = \binom{a}{n} e^{-n\mu t}(1 - e^{-\mu t})^{a-n}, \quad 0 \leqslant n \leqslant a. \tag{8.30}$$

In particular, $p_{0}(t) \to 1$ as $t \to \infty$.

The mean and variance of this distribution are

$$m(t) = ae^{-\mu t}, \tag{8.31}$$

and

$$\sigma^{2}(t) = ae^{-\mu t}(1 - e^{-\mu t}). \tag{8.32}$$

This time we see that the stochastic mean declines according to the negative exponential law we should expect of a deterministic model in which the loss in time Δt for a population of size $n(t)$ was precisely $\mu n \, \Delta t$.

We could if we wished generalize this simple death process in a manner analogous to the treatment of the general birth process in Sections 8.3 and 8.4. There is of course no possibility of an "explosion".

8.6 The simple birth-and-death process

Having worked out the implications of two simple models, one involving birth only, the other death only, we are now in a position to amalgamate both these aspects into a single model. We shall now assume that two types of transitions are possible, namely that the chance of any individual giving birth in time Δt is $\lambda \, \Delta t$, and the chance of dying in Δt is $\mu \, \Delta t$. It follows that in a population size $X(t)$ at time t the possible transitions may be classified as follows:

Chance of one birth $= \lambda X(t) \, \Delta t + o(\Delta t)$

Chance of one death $= \mu X(t) \, \Delta t + o(\Delta t)$

Chance of more than one of these events $= o(\Delta t)$

Chance of no change $= 1 - (\lambda + \mu)X(t)\,\Delta t + o(\Delta t)$

In the terminology of Section 7.4 we have transitions given by $j = +1$, -1; with $f_{+1}(X) = \lambda X$ and $f_{-1}(X) = \mu X$. The partial differential equation for the moment-generating function of the process can now be written down as

$$\frac{\partial M}{\partial t} = \{\lambda(e^{\theta} - 1) + \mu(e^{-\theta} - 1)\}\frac{\partial M}{\partial \theta}, \tag{8.33}$$

with initial condition $\qquad M(\theta, 0) = e^{a\theta}, \tag{8.34}$

if we start off with $X(0) = a$, as before.

The usual argument for the value of $p_n(t + \Delta t)$ easily leads to the differential-difference equations

$$\left.\begin{aligned}
\frac{dp_n}{dt} &= \lambda(n - 1)p_{n-1} - (\lambda + \mu)np_n + \mu(n + 1)p_{n+1}, \quad n \geqslant 1 \\[2mm]
\frac{dp_0}{dt} &= \mu p_1
\end{aligned}\right\}, \tag{8.35}$$

but it is simpler to proceed by solving equation (8.33) for the moment-generating function as follows.

Equation (8.33) is again of the standard linear type of partial differential equation discussed in Section 7.5. The subsidiary equations are

$$\frac{dt}{1} = \frac{-d\theta}{\lambda(e^{\theta} - 1) + \mu(e^{-\theta} - 1)} = \frac{dM}{0}. \tag{8.36}$$

The first and third expressions yield the integral

$$M = \text{constant}, \tag{8.37}$$

while the first and second give

$$dt = \frac{-e^{\theta}\,d\theta}{(e^{\theta} - 1)(\lambda e^{\theta} - \mu)},$$

from which we have by integration

$$t = \text{constant} - \frac{1}{\lambda - \mu}\log\frac{e^{\theta} - 1}{\lambda e^{\theta} - \mu}, \quad \mu \neq \lambda,$$

$$= \text{constant} + \frac{1}{\lambda(e^{\theta} - 1)}, \qquad \mu = \lambda.$$

The required form of the second intermediate integral is thus

$$
\left.
\begin{aligned}
\frac{(e^\theta - 1)e^{(\lambda-\mu)t}}{\lambda e^\theta - \mu} &= \text{constant}, \quad \mu \neq \lambda \\
\lambda t - \frac{1}{e^\theta - 1} &= \text{constant}, \quad \mu = \lambda
\end{aligned}
\right\}. \tag{8.38}
$$

For $\mu \neq \lambda$ the general solution of (8.33) can therefore be put in the form

$$
M = \Psi\left\{\frac{(e^\theta - 1)e^{(\lambda-\mu)t}}{\lambda e^\theta - \mu}\right\}. \tag{8.39}
$$

Using the boundary condition (8.34) gives

$$
e^{a\theta} = M(\theta, 0) = \Psi\left(\frac{e^\theta - 1}{\lambda e^\theta - \mu}\right). \tag{8.40}
$$

We now put $u = (e^\theta - 1)/(\lambda e^\theta - \mu)$, or

$$
e^\theta = \frac{\mu u - 1}{\lambda u - 1},
$$

in (8.40) to give the required function $\Psi(u)$, namely

$$
\Psi(u) = \left(\frac{\mu u - 1}{\lambda u - 1}\right)^a. \tag{8.41}
$$

From (8.41) and (8.39) we then have

$$
M(\theta, t) = \left(\frac{\mu v(\theta, t) - 1}{\lambda v(\theta, t) - 1}\right)^a, \tag{8.42}
$$

where

$$
v(\theta, t) = \frac{(e^\theta - 1)e^{(\lambda-\mu)t}}{\lambda e^\theta - \mu}. \tag{8.43}
$$

The probability-generating function is evidently

$$
P(x, t) = \left(\frac{\mu w(x, t) - 1}{\lambda w(x, t) - 1}\right)^a, \tag{8.44}
$$

where

$$
w(x, t) = \frac{(x - 1)e^{(\lambda-\mu)t}}{\lambda x - \mu}. \tag{8.45}
$$

In the special case $a = 1$, it is easy to expand the probability-generating function in (8.44) in powers of x^n. The required coefficient of x^n, i.e. $p_n(t)$, is easily found to be represented by

$$\left. \begin{array}{l} p_n(t) = (1 - \alpha)(1 - \beta)\beta^{n-1}, \quad n \geq 1 \\ p_0(t) = \alpha \end{array} \right\}, \tag{8.46}$$

where $\qquad \alpha = \dfrac{\mu(e^{(\lambda - \mu)t} - 1)}{\lambda e^{(\lambda - \mu)t} - \mu}, \qquad \beta = \dfrac{\lambda(e^{(\lambda - \mu)t} - 1)}{\lambda e^{(\lambda - \mu)t} - \mu}.$

We thus have a probability distribution which is given by the terms of a geometric series, except for the term corresponding to $n = 0$. When $a > 1$, the expressions for the probabilities are somewhat more complicated involving summations. Straightforward manipulations yield the result

$$\left. \begin{array}{l} p_n(t) = \displaystyle\sum_{j=0}^{\min(a,n)} \binom{a}{j} \binom{a + n - j - 1}{a - 1} \alpha^{a-j} \beta^{n-j}(1 - \alpha - \beta)^j \\ p_0(t) = \alpha^a \end{array} \right\}, \tag{8.47}$$

where α and β are as indicated in (8.46).

The stochastic mean and variance are most easily obtained for the general case $a \geq 1$ directly from (8.42). We simply write down the cumulant-generating function $K = \log M = a \log\{(\mu v - 1)/(\lambda v - 1)\}$ and pick out the coefficients of θ and θ^2. This gives

$$m(t) = ae^{(\lambda - \mu)t}, \tag{8.48}$$

and $\qquad \sigma^2(t) = \dfrac{a(\lambda + \mu)}{(\lambda - \mu)} e^{(\lambda - \mu)t}(e^{(\lambda - \mu)t} - 1). \tag{8.49}$

Once again the stochastic mean is equal to the corresponding deterministic value.

When $\mu = \lambda$ we can go back to the general form of the solution for $M(\theta, t)$ that arises when we use the alternative form of the second intermediate integral in (8.38). We then have

$$M = \Psi\left(\lambda t - \frac{1}{e^\theta - 1}\right), \tag{8.50}$$

leading to $\qquad M(\theta, t) = \left\{\dfrac{1 - (\lambda t - 1)(e^\theta - 1)}{1 - \lambda t(e^\theta - 1)}\right\}^a. \tag{8.51}$

The probability-generating function is thus

$$P(x, t) = \left\{\dfrac{1 - (\lambda t - 1)(x - 1)}{1 - \lambda t(x - 1)}\right\}^a. \tag{8.52}$$

In the special case when $a = 1$, we have

$$\left. \begin{aligned} p_n(t) &= \frac{(\lambda t)^{n-1}}{(1 + \lambda t)^{n+1}}, \quad n \geqslant 1 \\ p_0(t) &= \frac{\lambda t}{1 + \lambda t} \end{aligned} \right\}. \tag{8.53}$$

More generally, we use (8.51) to obtain the stochastic mean and variance for any a as

$$m(t) = a, \tag{8.54}$$

$$\sigma^2(t) = 2a\lambda t. \tag{8.55}$$

These results could of course have been obtained directly from (8.48) and (8.49) by letting $\mu \to \lambda$ and using l'Hôpital's rule where necessary. It follows from (8.54) and (8.55) that, in the case where the birth- and death-rates are exactly equal, the stochastic mean remains constant and equal to the initial population size, while the variance increases linearly with time.

Chance of extinction

We have already considered the chance of a process becoming extinct in the discussion of branching processes in Chapter 6 (see especially Section 6.2). Let us now look at this phenomenon in the present context. First, we note from (8.48) and (8.54) that for general a

$$\left. \begin{aligned} \lim_{t \to \infty} m(t) &= 0, \quad \lambda < \mu \\ &= a, \quad \lambda = \mu \\ &= \infty, \quad \lambda > \mu \end{aligned} \right\}. \tag{8.56}$$

Now, putting $x = 0$ in (8.44), we see that

$$p_0(t) = \left\{ \frac{\mu(e^{(\lambda-\mu)t} - 1)}{\lambda e^{(\lambda-\mu)t} - \mu} \right\}^a, \quad \lambda \neq \mu. \tag{8.57}$$

And from this last result, as the limiting form when $\mu = \lambda$, or directly from (8.52), we have

$$p_0(t) = \left(\frac{\lambda t}{\lambda t + 1} \right)^a, \quad \lambda = \mu. \tag{8.58}$$

If we now let $t \to \infty$ in (8.57) and (8.58) we obtain the following alternative values for the chance of extinction, namely

$$\left.\begin{aligned} \lim_{t \to \infty} p_0(t) &= 1, & \lambda \leqslant \mu \\ &= (\mu/\lambda)^a, & \lambda > \mu \end{aligned}\right\}. \tag{8.59}$$

Thus extinction of the process is certain unless the birth-rate *exceeds* the death-rate, and in the latter case the chance of extinction is $(\mu/\lambda)^a$.

It is of special interest to observe that even in the case when $\mu = \lambda$ and $m(t) \equiv a$, the chance of extinction is still unity. This curious result is explained by the fact that although a few populations will rise to very high values, most will be extinguished, in such a way as to achieve a constant mean. These results should emphasize the fallacy of attaching too much importance to stochastic mean values, even when these are exactly equal to the corresponding deterministic quantities.

It is perhaps worth remarking at this point that we have been fortunate in obtaining not only convenient, closed expressions for the probability-, moment-, and cumulant-generating functions, but have been able to derive explicit values for means and variances (and higher cumulants if desired) as well as for the individual probabilities. Frequently such explicit solutions are unavailable, and in the case of less tractable processes we must seek other means of exploring the properties of the model. One method of investigating the moments or cumulants is to obtain the appropriate partial differential equation, such as (8.33), which is usually easy, and equate coefficients of θ on both sides. This will yield a set of differential equations for the moments or cumulants, as the case may be. If we are lucky these will be soluble successively, giving, for example, $m(t)$, $\sigma^2(t)$, etc., in turn.

Thus in the case of the birth-and-death process we first write $M = e^K$ in (8.33) to give

$$\frac{\partial K}{\partial t} = \{\lambda(e^\theta - 1) + \mu(e^{-\theta} - 1)\} \frac{\partial K}{\partial \theta}. \tag{8.60}$$

If we now equate coefficients of θ on both sides of (8.60), using the usual expansion

$$K = \kappa_1 \theta + \kappa_2 \theta^2/2! + \cdots,$$

we obtain a set of differential equations for the cumulants, the first two of which are

$$\frac{d\kappa_1}{dt} = (\lambda - \mu)\kappa_1, \tag{8.61}$$

and

$$\frac{d\kappa_2}{dt} = 2(\lambda - \mu)\kappa_2 + (\lambda + \mu)\kappa_1. \tag{8.62}$$

The first equation (8.61) integrates immediately to give the result already obtained in (8.48). Substituting this value of κ_1 in (8.62) then yields an equation for κ_2 which also integrates very readily to supply the result in (8.49).

This method often works well for processes in which the transition probabilities are only linear functions of the random variable. But when non-linear functions are involved, as in epidemic models, we may find that the first equation involves κ_2 as well as κ_1, the second involves κ_3 as well as κ_2 (and perhaps κ_1), and so on. This happens with epidemics (see Chapter 12) and competition models (see Chapter 13). The method then fails, since we cannot solve the first equation to get the process of successive solutions started. However, it may be possible in such circumstances to adopt an approximation, in which all cumulants beyond a certain point are neglected (see Section 15.4).

8.7 The effect of immigration

In many biological populations some form of migration is an essential characteristic. Consequently, we should devote some attention to introducing this phenomenon into the birth-and-death process just discussed. So far as the emigration of individuals out of the population is concerned, it is clear that this can be allowed for by a suitable adjustment of the death-rate, since, for deaths and emigrations, we can take the chance of a single loss in time Δt as proportional to $X(t)\,\Delta t$. With immigrations, on the other hand, the situation is different, for the simplest reasonable assumption about immigration is that it occurs as a random Poisson process, independent of population size.

Suppose, therefore, that we consider the birth-and-death process defined at the beginning of Section 8.6, plus a random accession of immigrants, with immigration-rate ν. This means that the chance of an increase in population of one unit in time Δt is now $\lambda X(t)\,\Delta t + \nu\,\Delta t$, while the chance of a loss of one unit is still $\mu X(t)\,\Delta t$. In the terminology of Section 7.4, we have $f_{+1}(X) = \lambda X + \nu$, and $f_{-1}(X) = \mu X$. The partial differential equation for the moment-generating function is accordingly

$$
\begin{aligned}
\frac{\partial M}{\partial t} &= (e^{\theta} - 1)\left(\lambda\frac{\partial}{\partial\theta} + \nu\right)M + \mu(e^{-\theta} - 1)\frac{\partial M}{\partial\theta} \\
&= \{\lambda(e^{\theta} - 1) + \mu(e^{-\theta} - 1)\}\frac{\partial M}{\partial\theta} + \nu(e^{\theta} - 1)M.
\end{aligned}
\tag{8.63}
$$

This differs from (8.33) only in the addition of the last term on the right, which is clearly a Poisson process component.

We can solve (8.63) without too much difficulty. The subsidiary equations are

$$\frac{dt}{1} = \frac{-d\theta}{\lambda(e^\theta - 1) + \mu(e^{-\theta} - 1)} = \frac{dM}{v(e^\theta - 1)M}. \tag{8.64}$$

The first two elements of (8.64) are precisely the same as those occurring in (8.36), and therefore yield the integral given in the first line of (8.38), i.e.

$$\frac{(e^\theta - 1)e^{(\lambda - \mu)t}}{\lambda e^\theta - \mu} = \text{constant.} \tag{8.65}$$

The second and third elements of (8.64) give

$$\frac{dM}{M} = \frac{-vd\theta}{\lambda - \mu e^{-\theta}}$$

$$= \frac{-ve^\theta \, d\theta}{\lambda e^\theta - \mu}, \tag{8.66}$$

from which we obtain

$$\log M = -\frac{v}{\lambda} \log(\lambda e^\theta - \mu) + \text{constant,}$$

or $\qquad\qquad (\lambda e^\theta - \mu)^{v/\lambda} M = \text{constant.} \tag{8.67}$

The required general solution of (8.63) is thus of the form

$$(\lambda e^\theta - \mu)^{v/\lambda} M = \Psi\left\{ \frac{(e^\theta - 1)e^{(\lambda - \mu)t}}{\lambda e^\theta - \mu} \right\}. \tag{8.68}$$

If we choose the initial condition $X(0) = a$, substitution in (8.68) of $t = 0$, $M(\theta, 0) = e^{a\theta}$, gives

$$(\lambda e^\theta - \mu)^{v/\lambda} e^{a\theta} = \Psi\left(\frac{e^\theta - 1}{\lambda e^\theta - \mu} \right). \tag{8.69}$$

As before, we put $u = (e^\theta - 1)/(\lambda e^\theta - \mu)$, i.e.

$$e^\theta = \frac{\mu u - 1}{\lambda u - 1},$$

in (8.69) to give

$$\Psi(u) = \left(\frac{\mu u - 1}{\lambda u - 1} \right)^a \left(\frac{\mu - \lambda}{\lambda u - 1} \right)^{v/\lambda}. \tag{8.70}$$

From (8.70) and (8.68) we obtain the required function $M(\theta, t)$ as

$$M(\theta, t) = \frac{(\lambda - \mu)^{\nu/\lambda}\{\mu(e^{(\lambda - \mu)t} - 1) - (\mu e^{(\lambda - \mu)t} - \lambda)e^{\theta}\}^{a}}{\{(\lambda e^{(\lambda - \mu)t} - \mu) - \lambda(e^{(\lambda - \mu)t} - 1)e^{\theta}\}^{a + \nu/\lambda}}. \qquad (8.71)$$

This general form is a little complicated, but it is interesting to examine a few special cases.

The case $a = 0$

Let us first consider the case $a = 0$. The population starts off from zero, but is of course raised to a positive size in the first instance by immigration. In a similar way the process does not become extinct if all the individuals should die, since it is restarted by the introduction of new individuals from outside. If we put $a = 0$ and $e^{\theta} = x$ in (8.71), we have the appropriate probability-generating function

$$P(x, t) = \left(\frac{\lambda - \mu}{\lambda e^{(\lambda - \mu)t} - \mu}\right)^{\nu/\lambda} \left\{1 - \frac{\lambda(e^{(\lambda - \mu)t} - 1)}{\lambda e^{(\lambda - \mu)t} - \mu} x\right\}^{-\nu/\lambda}. \qquad (8.72)$$

The population thus has a negative binomial distribution with mean

$$m(t) = \frac{\nu}{\lambda - \mu} (e^{(\lambda - \mu)t} - 1). \qquad (8.73)$$

For $\lambda > \mu$, the expected population size grows exponentially for large t, with rate $\lambda - \mu$; while for $\lambda < \mu$ the limiting value for large t is clearly $\nu/(\mu - \lambda)$. In the intermediate case where $\mu = \lambda$, we see from (8.73) that $m(t) = \nu t$ for all t. The limiting properties as $t \to \infty$ are thus

$$\left.\begin{aligned} m(t) &\sim \frac{\nu}{\lambda - \mu} e^{(\lambda - \mu)t}, & \lambda > \mu \\ &= \nu t, & \lambda = \mu \\ &\sim \frac{\nu}{\mu - \lambda}, & \lambda < \mu \end{aligned}\right\}. \qquad (8.74)$$

The case $\lambda < \mu$ is of special interest, and worth examining in more detail. As $t \to \infty$, $e^{(\lambda - \mu)t} \to 0$. Substituting the latter value in (8.72) gives the complete limiting distribution

$$P(x, \infty) = \left(\frac{\mu - \lambda x}{\mu - \lambda}\right)^{-\nu/\lambda}, \quad \lambda < \mu. \qquad (8.75)$$

We thus have a limiting stable distribution, but the precise value of the exponent depends on the intensity of immigration.

The behavior of a process involving only death and immigration can also be deduced directly from (8.75) by considering the limit as $\lambda \to 0$. The limiting form of the probability-generating function is

$$\lim_{\lambda \to 0} P(x, \infty) = \lim_{\lambda \to 0} \left\{1 - \frac{\lambda}{\mu}(x - 1) + O\left(\frac{\lambda}{\mu}\right)^2\right\}^{-\nu/\lambda}$$

$$= \exp\left\{\frac{\nu}{\mu}(x - 1)\right\}, \tag{8.76}$$

which represents a Poisson distribution with parameter ν/μ. Such a model may be useful in practice to represent the fluctuations in the number of particles in a small volume under direct observation. Immigration from the surrounding medium may be regarded as a random Poisson process with constant parameter ν, while emigration or death is taken to have a rate proportional to the population size. This representation has been used in the study of colloidal particles in suspension by Chandrasekhar (1943), and to investigate the movements of spermatozoa by Rothschild (1953).

The case $\nu \to 0$

Let us now see what happens if ν is very small, and in the limit we let $\nu \to 0$, keeping λ constant. This means we assume just sufficient immigration to restart the process if it should happen to become extinct. Suppose we put $\nu/\lambda = \gamma$ in (8.72). Then we can write the probability-generating function as

$$P(x, t) = A^\gamma(1 - Bx)^{-\gamma}, \tag{8.77}$$

where

$$A = \frac{\lambda - \mu}{\lambda e^{(\lambda - \mu)t} - \mu}, \qquad B = \frac{\lambda(e^{(\lambda - \mu)t} - 1)}{\lambda e^{(\lambda - \mu)t} - \mu}. \tag{8.78}$$

In some experimental situations the zero class is strictly unobservable. This occurs, for example, if we consider the numbers of specimens of different species caught in a trap: no species can be represented by zero individuals. We are then interested in the distribution conditional on the zero class being absent. Equation (8.77) then yields a probability-generating function $P_1(x, t)$ given by

$$P_1(x, t) \propto Bx + \frac{\gamma + 1}{2!}(Bx)^2 + \frac{(\gamma + 1)(\gamma + 2)}{3!}(Bx)^3 + \cdots, \tag{8.79}$$

where we have removed the constant term in (8.77), and have also taken out a factor γA^γ. Now

$$\lim_{\gamma \to 0} P_1(x, t) \propto -\log(1 - Bx),$$

and since $P_1(1, t) = 1$, the divisor required on the right is $-\log(1 - B)$. We therefore have the logarithmic distribution given by

$$P_1(x, t) = \frac{\log(1 - Bx)}{\log(1 - B)}. \tag{8.80}$$

8.8 The general birth-and-death process

In Section 8.3 we generalized the treatment of the simple birth process discussed in Section 8.2 by assuming that the chance of a birth in the interval Δt was $\lambda_n \Delta t$ where n was the population size at time t. In a similar way we can generalize the model for the simple birth-and-death process by adding to the foregoing the further assumption that the chance of a death in Δt is $\mu_n \Delta t$. The basic differential-difference equations then come out to be

$$\left. \begin{aligned} \frac{dp_n}{dt} &= \lambda_{n-1}p_{n-1} - (\lambda_n + \mu_n)p_n + \mu_{n+1}p_{n+1}, \quad n \geqslant 1 \\ \frac{dp_0}{dt} &= -\lambda_0 p_0 + \mu_1 p_1 \end{aligned} \right\}. \tag{8.81}$$

It is beyond the scope of the present book to discuss the solution of (8.81) in detail. But it can be shown that for arbitrary coefficients $\lambda_n \geqslant 0$, $\mu_n \geqslant 0$, there always exists a positive solution $p_n(t)$ such that $\sum_n p_n(t) \leqslant 1$. If, moreover, the coefficients are bounded, or increase sufficiently slowly this solution will be unique and $\sum_n p_n(t) = 1$. Although there is considerable theoretical interest in the possibility of $\sum_n p_n(t) < 1$, when there may be an infinity of solutions, in practice we can usually assume with safety that the uniqueness condition is satisfied.

There are several theorems relating to the conditions required of the λ_n and μ_n if the process is not to be explosive (see Bharucha-Reid, 1960, p. 91 et seq.), and in certain cases it might be useful to test a proposed model to ensure that it did not entail undesirable properties. In the present discussion we merely remark that the following conditions are sufficient for a unique solution:
(a) $\lambda_n = 0$ for some

$$n \geqslant 1. \tag{8.82}$$

This means that an upper limit of n is automatically placed on the population size.

(b) $\lambda_n > 0$ for $n \geqslant a$, and

$$\sum_{n=a}^{\infty} \lambda_n^{-1} = \infty. \tag{8.83}$$

This restricts the magnitude of the λ_n to ensure that an infinity of births cannot occur in a finite time (compare the discussion of Section 8.4).

(c) $\lambda_n > 0$ for $n \geqslant a$, and

$$\sum_{n=a}^{\infty} \frac{\mu_n \mu_{n-1} \cdots \mu_a}{\lambda_n \lambda_{n-1} \cdots \lambda_a} = \infty. \tag{8.84}$$

This ensures the finiteness of the population by insisting on a certain balance of birth- and death-rates.

(d) $\lambda_n > 0$ for $n \geqslant a$, and

where

$$\left. \begin{array}{c} \displaystyle\sum_{n=a}^{\infty} w_n = \infty \\[2mm] w_n = \dfrac{1}{\lambda_n} + \dfrac{\mu_n}{\lambda_n \lambda_{n-1}} + \cdots + \dfrac{\mu_n \cdots \mu_{a+1}}{\lambda_n \cdots \lambda_a} + \dfrac{\mu_n \cdots \mu_a}{\lambda_n \cdots \lambda_a} \end{array} \right\}. \tag{8.85}$$

Here both kinds of factors in (b) and (c) are combined.

In the simplest birth-and-death process for which λ_n and μ_n are each proportional to n, it is obvious that $\sum \lambda_n^{-1} \propto 1 + \frac{1}{2} + \frac{1}{3} + \cdots \to \infty$. Thus we can apply condition (b) above and the solution for finite t is non-explosive. Notice, however, that if $\lambda > \mu$, then referring to (8.44) we see that

$$\lim_{t \to \infty} P(x, t) = (\mu/\lambda)^a < 1. \tag{8.86}$$

Thus the probability not absorbed at the value zero eventually moves to infinity, and the process is still what may be called dissipative.

8.9 Multiplicative processes

In Chapter 6 we studied a special type of discrete time Markov chain, namely a discrete branching process in which the individuals of each generation reproduced independently of one another. This idea can readily be extended to a continuous time situation. It is also convenient to suppose that the individuals alive at any given time each give rise to an independent birth-and-death process, i.e. that the whole process is multiplicative.

It is clear that the simple birth-and-death process already discussed is multiplicative, though this may not be so in general. For example, the multiplicative property may be lost if the birth- or death-rates per individual are not constant, but depend on the total population size.

In the discrete time model we wrote $P(x)$ for the probability-generating function of the number of offspring of any individual, and $P_n(x)$ for the probability-generating function of the distribution of the total population size at the nth generation. We obtained in (6.7) the dual forms of a basic functional relationship, repeated here for convenience, namely

$$P_n(x) = P_{n-1}(P(x)) = P(P_{n-1}(x)). \tag{8.87}$$

The continuous time analog is clearly, for small Δt,

$$P(x, t + \Delta t) = P(f(x, \Delta t), t) = f(P(x, t), \Delta t), \tag{8.88}$$

where as usual we write $P(x, t)$ for the probability-generating function of population size at time t, and use $f(x, \Delta t)$ for the probability-generating function of the transitions occurring over Δt for a *single* individual. It is reasonable to suppose that $f(x, \Delta t)$ has the form

$$f(x, \Delta t) = x + g(x)\,\Delta t + o(\Delta t), \tag{8.89}$$

in which the term x is included on the right since we must be left with the individual being considered as $\Delta t \to 0$.

Let us now substitute (8.89) into the equation given by taking the first two items in (8.88). We obtain

$$P(x, t + \Delta t) = P(x + g(x)\,\Delta t, t)$$

$$\doteq P(x, t) + g(x)\,\Delta t\, \frac{\partial P(x, t)}{\partial x},$$

from which it follows that

$$\frac{\partial P(x, t)}{\partial t} = g(x)\, \frac{\partial P(x, t)}{\partial x}. \tag{8.90}$$

In a similar manner the first and third items of (8.88) yield

$$P(x, t + \Delta t) \doteq P(x, t) + g(P(x, t))\,\Delta t.$$

Hence

$$\frac{\partial P(x, t)}{\partial t} = g(P(x, t)). \tag{8.91}$$

The two equations (8.90) and (8.91) clearly correspond to the forward and backward equations of our earlier discussions.

Let us apply these results directly to the simple birth-and-death process, which is easily seen to be of a multiplicative character. The probability-generating function $f(x, \Delta t)$ is evidently

$$
\begin{aligned}
f(x, \Delta t) &= \lambda \Delta t x^2 + \mu \Delta t + (1 - \lambda \Delta t - \mu \Delta t)x \\
&= x + \{\lambda(x^2 - x) + \mu(1 - x)\} \Delta t,
\end{aligned}
\tag{8.92}
$$

since, on the definitions at the beginning of Section (8.6), the probability of a birth in time Δt is $\lambda \Delta t$ and this gives us two individuals in all, while the chance of the single individual disappearing is $\mu \Delta t$. The function $g(x)$ is thus given by

$$
g(x) = \lambda(x^2 - x) + \mu(1 - x).
\tag{8.93}
$$

The forward equation (8.90) is therefore

$$
\frac{\partial P}{\partial t} = \{\lambda(x^2 - x) + \mu(1 - x)\} \frac{\partial P}{\partial x},
\tag{8.94}
$$

corresponding to (8.33) with $x = e^\theta$, and the backward equation (8.91) is

$$
\frac{\partial P}{\partial t} = (P - 1)(\lambda P - \mu).
\tag{8.95}
$$

The backward equation thus gives us a method of representation which we have not used so far in the treatment of any specific process (although we have discussed it in the general theory of Section 7.6). Since the partial differential equation in (8.91) involves only one partial differential coefficient, we can write it in the form

$$
dt = \frac{dP}{g(P)},
$$

and integrate directly to give

$$
t = \int_{P(x,0)}^{P(x,t)} \frac{du}{g(u)}.
\tag{8.96}
$$

If we consider, for the moment, only processes starting at $t = 0$ with a single individual, we shall have a lower limit of $P(x, 0) = x$ in (8.96).

Now let us use the form of $g(x)$ given in (8.93). The integral in (8.96) can be evaluated immediately as

$$
\begin{aligned}
t &= \int_x^P \frac{du}{(u - 1)(\lambda u - \mu)} \\
&= \left[\frac{1}{\lambda - \mu} \log \frac{u - 1}{\lambda u - \mu} \right]_x^P.
\end{aligned}
$$

Hence
$$P(x, t) = \frac{\mu w(x, t) - 1}{\lambda w(x, t) - 1},$$ (8.97)

where
$$w(x, t) = \frac{(x - 1)e^{(\lambda - \mu)t}}{\lambda x - \mu}.$$ (8.98)

These results are precisely those already given in (8.44) and (8.45) for the special case $a = 1$. But since we are dealing with a multiplicative process the probability-generating function at time t for a process starting with a individuals is $\{P(x, t)\}^a$, which is the general result appearing in (8.44).

It will be noticed that the solution of the backward equation is obtained with appreciably less effort than is required for the forward equation, previously used at the beginning of Section 8.6. The possibility of such a simplification should always be borne in mind.

PROBLEMS FOR SOLUTION

1. Consider a population with death-rate μ, which produces no new individuals (i.e. the birth-rate $\lambda = 0$), but which is maintained at least partially by mmigration occurring as a Poisson process with parameter ν. Initially there are a individuals.

Write down the partial differential equation for the moment-generating function, and solve it. What is the explicit form of the probability-generating function?

What happens as $t \to \infty$?

[This is the special case of the model discussed in Section 8.7 when $\lambda = 0$. It is suggested that the problem should be worked from first principles as an exercise.]

2. Calculate the stochastic mean $m(t)$ for the process investigated in Problem 1.

3. A group of N animals graze in a field. The animals graze independently. Periods of grazing and resting for any one animal alternate, having negative exponential distributions with parameters μ and λ, respectively.

Write down the partial differential equation for the number of animals grazing at time t.

4. What is the steady-state solution of Problem 3 as $t \to \infty$?

5. Calculate the mean number of animals grazing in Problem 3 at time t if initially there are a when $t = 0$.

6. Consider the general birth-and-death process of Section 8.8 with birth-rate λ_n and death-rate μ_n for a population of size n. Assume that λ_n, $\mu_n \neq 0$ for $n > 0$ and λ_0, $\mu_0 = 0$. Let $q_a(t)$ be the probability that $n = 0$ at time t given that $n = a$ at $t = 0$. Show that $q_a(t)$ satisfies

$$\frac{dq_a(t)}{dt} = \lambda_n\{q_{a+1}(t) - q_a(t)\} + \mu_n\{q_{a-1}(t) - q_a(t)\}, \quad a = 1, 2, \cdots.$$

Hence obtain the condition that the λ_n and μ_n must satisfy for $q_a(t)$ to tend to unity as t increase.

7. Write down the Kolmogorov forward and backward equations for the probabilities $p_{ij}(t)$ for

   (i) the general birth process;

   (ii) the general birth-and-death process.

CHAPTER 9

Some Non-Homogeneous Processes

9.1 Introduction

In Chapter 8 we confined the discussion to processes which were homogeneous with respect to time, that is the transition probabilities were at most functions of the population size and did not involve any arbitrary dependence on the time variable. It is interesting and instructive to investigate the complications that arise if we relax this restriction. For example, we may wish to examine the properties of a birth-and-death model, of the type dealt with in Section 8.6, where the birth- and death-rates per individual are taken to be functions of time, say $\lambda(t)$ and $\mu(t)$, respectively. In particular, this formulation is sometimes of value as an approximation to a very much more complicated situation (see, for example, the application to the prey-predator problem in Section 13.3). First, however, we shall discuss a special process which happens to yield a fairly simple solution, using the standard method of handling linear partial differential equations.

9.2 The Pólya process

Let us consider the type of problem arising in the treatment of electron–photon cascade theory. Here we are primarily concerned with a population of physical particles, whose total number at time t is given by the random variable $X(t)$. The actual processes are usually very complicated, and various simplifying approximations have been used (see Bharucha–Reid, 1960, Section 5.2, for references). One of these, the Pólya process, assumes a non-homogeneous birth process in which the chance of a new particle appearing in a population of size $X(t)$ during the interval Δt is

$$\frac{\lambda\{1 + \mu X(t)\}}{1 + \lambda \mu t} \Delta t, \tag{9.1}$$

where λ and μ are suitable non-negative constants. It is clear from the form of (9.1) that we can regard a new particle as arising from two sources, one involving the random Poisson type of birth with parameter $\lambda(1 + \lambda\mu t)^{-1}$, the other being a simple birth type of process with a birth-rate of $\lambda\mu(1 + \lambda\mu t)^{-1}$ per particle. Note, however, the introduction of the decay factor $(1 + \lambda\mu t)^{-1}$.

Whether such assumptions are likely to be valid for any population of biological organisms is uncertain. Nevertheless, the Pólya process is a good example of a mathematically tractable, non-homogeneous stochastic process, and therefore seems worth examining in a little detail.

It is easy to apply the random-variable technique of Section 7.4, which we have previously used only for homogeneous processes. There is nothing in the derivation of equation (7.28), for example, which restricts us to transition functions f_j which are functions of X only. We thus have only one type of transition given by $j = +1$, and from (9.1) we can write

$$f_{+1} = \frac{\lambda(1 + \mu X)}{1 + \lambda\mu t}. \tag{9.2}$$

Hence the basic partial differential equation in (7.28) becomes

$$\frac{\partial M}{\partial t} = \frac{\lambda(e^\theta - 1)}{1 + \lambda\mu t}\left(M + \mu\frac{\partial M}{\partial\theta}\right), \tag{9.3}$$

or, writing $K = \log M$, we have the corresponding equation for the cumulant-generating function

$$\frac{\partial K}{\partial t} = \frac{\lambda(e^\theta - 1)}{1 + \lambda\mu t}\left(1 + \mu\frac{\partial K}{\partial\theta}\right). \tag{9.4}$$

It is obvious from the form of (9.4) that simple transformations of the variables would make the subsequent algebra easier. Let us use

$$\lambda\mu t = T, \qquad e^\theta - 1 = \phi, \qquad \mu K = L. \tag{9.5}$$

With these quantities substituted in (9.4), we obtain

$$(1 + T)\frac{\partial L}{\partial T} - \phi(1 + \phi)\frac{\partial L}{\partial\phi} = \phi. \tag{9.6}$$

The subsidiary equations are

$$\frac{dT}{1 + T} = -\frac{d\phi}{\phi(1 + \phi)} = \frac{dL}{\phi}. \tag{9.7}$$

The first two expressions in (9.7) integrate to give

$$\log(1 + T) = \log \frac{1 + \phi}{\phi} + \text{constant},$$

or $\qquad\qquad \frac{(1 + T)\phi}{1 + \phi} = \text{constant}. \qquad\qquad (9.8)$

Next, the second and third items in (9.7) integrate to give

$$e^L(1 + \phi) = \text{constant}. \qquad\qquad (9.9)$$

Using the two independent integrals in (9.8) and (9.9) we can write the general solution of (9.6) as

$$e^L(1 + \phi) = \Psi\left\{\frac{(1 + T)\phi}{1 + \phi}\right\}, \qquad\qquad (9.10)$$

where Ψ is a suitable arbitrary function.

Suppose the process starts, as it can, at $t = 0$ with $X(0) = 0$. Now when $t = 0$, we have $T = K = L = 0$. Thus (9.10) becomes

$$1 + \phi = \Psi\left(\frac{\phi}{1 + \phi}\right). \qquad\qquad (9.11)$$

Now put $\phi/(1 + \phi) = u$, i.e. $1 + \phi = (1 - u)^{-1}$, in (9.11), and we obtain the form of Ψ as

$$\Psi(u) = \frac{1}{1 - u}. \qquad\qquad (9.12)$$

Applying this result to (9.10) yields

$$e^L(1 + \phi) = \frac{1 + \phi}{1 - T\phi},$$

or $\qquad\qquad L = -\log(1 - T\phi),$

which, on substituting back from (9.5), gives the required solution

$$K(\theta, t) = -\mu^{-1} \log\{1 - \lambda\mu t(e^\theta - 1)\}. \qquad\qquad (9.13)$$

The moment-generating function is thus

$$M(\theta, t) = \{1 - \lambda\mu t(e^\theta - 1)\}^{-1/\mu}, \qquad\qquad (9.14)$$

and the probability-generating function is

$$P(x, t) = \{(1 + \lambda\mu t) - \lambda\mu tx\}^{-1/\mu}, \qquad\qquad (9.15)$$

which is yet another derivation of a negative binomial distribution (see also Sections 8.2 and 8.7). This time the individual probabilities are

$$p_n(t) = \frac{(\lambda t)^n}{n!} (1 + \lambda\mu t)^{-n-1/\mu} \prod_{j=1}^{n-1} (1 + j\mu), \quad n = 1, 2, \cdots. \tag{9.16}$$

The stochastic mean and variance can be obtained in the usual way directly from the cumulant-generating function in (9.13). Alternatively, we can use the known values for a negative binomial with the parameters appearing in (9.15). Hence

$$m(t) = \lambda t, \tag{9.17}$$

$$\sigma^2(t) = \lambda t(1 + \lambda\mu t). \tag{9.18}$$

Thus the mean increases linearly with time, while the variance has a Poisson value for very small t but increases steadily above this level.

It is not difficult, if desired, to obtain explicitly the more general result for the initial condition of a particles present at time $t = 0$. This time we should simply have $\Psi(u) = (1 - u)^{-1-a}$.

In describing some actual processes with births but no deaths the Pólya model offers a little more scope than either the Poisson process or the Yule–Furry type of simple birth process, as it involves two parameters, λ and μ, instead of only one. On the other hand it does commit one to a particular kind of non-homogeneity in time.

9.3 A simple non-homogeneous birth-and-death process

Let us now turn to the simplest kind of non-homogeneous birth-and-death process. We could, of course, examine the cases of birth processes and death processes separately, but the more general model turns out to be fairly manageable. We consider the type of model already investigated in Section 8.6, with the single modification that the birth- and death-rates are to be taken as functions of the time, e.g. $\lambda(t)$ and $\mu(t)$, respectively. This modification leaves both the general form of the basic differential-difference equations for the probabilities in (8.35), and the partial differential equation for the moment-generating function in (8.33), unaltered. Working with the equation for the probability-generating function, i.e. putting $e^\theta = x$ in (8.33), we can thus write

$$\frac{\partial P}{\partial t} = (x - 1)\{\lambda(t)x - \mu(t)\} \frac{\partial P}{\partial x}, \tag{9.19}$$

with initial condition

$$P(x, 0) = x^a. \tag{9.20}$$

The equation (9.19) yields quite easily to the standard method. Thus we first write down the subsidiary equations

$$\frac{dt}{1} = \frac{dx}{(x - 1)(\mu - \lambda x)} = \frac{dP}{0}, \tag{9.21}$$

remembering that now both λ and μ are functions of t. The first and third items immediately give one integral

$$P = \text{constant.} \tag{9.22}$$

Another integral can be found from the solution of

$$\frac{dx}{dt} = (x - 1)(\mu - \lambda x), \tag{9.23}$$

which simplifies somewhat by use of the substitution

$$x - 1 = \frac{1}{y}. \tag{9.24}$$

We find

$$\frac{dy}{dt} = \lambda - (\mu - \lambda)y, \tag{9.25}$$

which now involves only a linear function of y on the right. The usual integrating factor for such an equation is $e^{\rho(t)}$, where

$$\rho(t) = \int_0^t \{\mu(\tau) - \lambda(\tau)\} \, d\tau. \tag{9.26}$$

We can thus write

$$e^\rho \frac{dy}{dt} + (\mu - \lambda)e^\rho y = \lambda e^\rho,$$

which, on integration with respect to t, yields

$$ye^\rho = \int_0^t \lambda(\tau)e^{\rho(\tau)} \, d\tau + \text{constant},$$

or

$$\frac{e^{\rho(t)}}{x - 1} - \int_0^t \lambda(\tau)e^{\rho(\tau)} \, d\tau = \text{constant}, \tag{9.27}$$

which is the second integral required.

The general solution of (9.19) can therefore be expressed in the form

$$P(x, t) = \Psi\left(\frac{e^\rho}{x - 1} - \int_0^t \lambda e^\rho \, d\tau\right), \tag{9.28}$$

where the arguments of λ and ρ have been suppressed for convenience. Putting $t = 0$ in (9.28), and using the initial condition (9.20), gives

$$x^a = \Psi\left(\frac{1}{x-1}\right). \tag{9.29}$$

Writing $u = (x - 1)^{-1}$, or $x = 1 + u^{-1}$, gives the desired form of Ψ as

$$\Psi(u) = \left(1 + \frac{1}{u}\right)^a. \tag{9.30}$$

The final expression for the probability-generating function is therefore given by applying (9.30) to (9.28), i.e.

$$P(x, t) = \left\{1 + \frac{1}{\dfrac{e^{\rho(t)}}{x-1} - \displaystyle\int_0^t \lambda(\tau)e^{\rho(\tau)}\,d\tau}\right\}^a, \tag{9.31}$$

where $\rho(t)$ is defined by (9.26).

There is little difficulty in expanding (9.31) in powers of x to obtain the individual state probabilities. The general formula is a little complicated, but can be written as

$$\left.\begin{array}{l} p_n(t) = \displaystyle\sum_{j=0}^{min(a,n)} \binom{a}{j}\binom{a+n-j-1}{a-1}\alpha^{a-j}\beta^{n-j}(1-\alpha-\beta)^j \\[2ex] p_0(t) = \alpha^a \end{array}\right\}, \tag{9.32}$$

where

$$\alpha = 1 - \frac{1}{e^{\rho(t)} + A(t)}, \qquad \beta = 1 - \frac{e^{\rho(t)}}{e^{\rho(t)} + A(t)},$$

and

$$A(t) = \int_0^t \lambda(\tau)e^{\rho(\tau)}\,d\tau.$$

In the special case of a homogeneous process with $\lambda(t) = \lambda$ and $\mu(t) = \mu$, these expressions simplify to the formulas previously given in (8.47).

Chance of extinction

The chance of the process becoming extinct by time t is, as usual, given by $p_0(t)$. From (9.32) this is

$$p_0(t) = \left\{1 - \frac{1}{e^{\rho(t)} + A(t)}\right\}^a. \tag{9.33}$$

Now

$$e^{\rho(t)} + A(t) = e^{\rho(t)} + \int_0^t \lambda(\tau)e^{\rho(\tau)} \, d\tau$$

$$= e^{\rho(t)} + \int_0^t \mu(\tau)e^{\rho(\tau)} \, d\tau - \int_0^t \{\mu(\tau) - \lambda(\tau)\}e^{\rho(\tau)} \, d\tau$$

$$= e^{\rho(t)} + \int_0^t \mu(\tau)e^{\rho(\tau)} \, d\tau - [e^{\rho(\tau)}]_0^t, \quad \text{using (9.26),}$$

$$= 1 + \int_0^t \mu(\tau)e^{\rho(\tau)} \, d\tau.$$

Thus substituting this in (9.33) gives

$$p_0(t) = \left\{ \frac{\displaystyle\int_0^t \mu(\tau)e^{\rho(\tau)} \, d\tau}{1 + \displaystyle\int_0^t \mu(\tau)e^{\rho(\tau)} \, d\tau} \right\}^a \tag{9.34}$$

It is evident from (9.34) that the chance of extinction tends to unity as $t \to \infty$, if and only if

$$\lim_{t \to \infty} \int_0^t \mu(\tau)e^{\rho(\tau)} \, d\tau = \infty. \tag{9.35}$$

A particular application of these results is made in Section 13.3.

Calculation of cumulants

From the expression for the probability-generating function in (9.31) we can easily write down the cumulant-generating function as

$$K(\theta, t) = \log P(e^\theta, t),$$

and expand in powers of θ to find the cumulants. Actually, it is slightly easier in the present context to obtain the first two cumulants by using the appropriate partial differential equation for the cumulant-generating function and equating coefficients of θ on both sides. This gives two simple ordinary differential equations to solve. Thus from (9.19), writing $x = e^\theta$, we have

$$\frac{\partial M}{\partial t} = \{\lambda(t)(e^\theta - 1) + \mu(t)(e^{-\theta} - 1)\} \frac{\partial M}{\partial \theta},$$

from which we obtain

$$\frac{\partial K}{\partial t} = \{\lambda(t)(e^{\theta} - 1) + \mu(t)(e^{-\theta} - 1)\} \frac{\partial K}{\partial \theta}. \tag{9.36}$$

Equating coefficients of θ and θ^2 on both sides of (9.36) yields the two equations

$$\frac{d\kappa_1}{dt} = \{\lambda(t) - \mu(t)\}\kappa_1, \tag{9.37}$$

and $$\frac{d\kappa_2}{dt} = \{\lambda(t) + \mu(t)\}\kappa_1 + 2\{\lambda(t) - \mu(t)\}\kappa_2, \tag{9.38}$$

with the initial conditions

$$\kappa_1(0) = a, \quad \kappa_2(0) = 0. \tag{9.39}$$

Equation (9.37) integrates directly, after multiplying through by the integrating factor $e^{\rho(t)}$, to give

$$\kappa_1 \equiv m(t) = ae^{-\rho(t)}. \tag{9.40}$$

We now substitute this value of κ_1 in (9.38), multiply through by the integrating factor $e^{2\rho(t)}$, and integrate, obtaining

$$\kappa_2 \equiv \sigma^2(t) = ae^{-2\rho(t)} \int_0^t \{\lambda(\tau) + \mu(\tau)\}e^{\rho(\tau)}\, d\tau. \tag{9.41}$$

An application of these results to cascade showers in cosmic ray theory has been made by Arley (1943). He adopted a simplified model in which the birth-rate was constant, i.e. $\lambda(t) \equiv \lambda$, but the death-rate increased linearly with time, i.e. $\mu(t) = \mu t$.

In this case we have the function $\rho(t)$ given by

$$\rho(t) = \int_0^t \{\mu(\tau) - \lambda(\tau)\}\, d\tau$$

$$= \tfrac{1}{2}\mu t^2 - \lambda t. \tag{9.42}$$

The mean and variance, given by (9.40) and (9.41), are easily calculated to be

$$m(t) = e^{\lambda t - \frac{1}{2}\mu t^2}, \tag{9.43}$$

and

$$\sigma^2(t) = m(t)\{1 - m(t)\} + 2\lambda\{m(t)\}^2 \int_0^t \frac{d\tau}{m(\tau)}, \tag{9.44}$$

where we assume a single particle initially, i.e. $a = 1$.

The integral in (9.35) has the limit

$$\lim_{t \to \infty} \int_0^t \mu\tau e^{\frac{1}{2}\mu\tau^2 - \lambda\tau} \, d\tau = \infty, \tag{9.45}$$

so that extinction is certain.

9.4 The effect of immigration

In general the effects of immigration can be investigated by introducing an immigration-rate $v(t)$ which is a function of the time. We then have an equation for the moment-generating function which is similar to (8.63), except that now λ, μ, and v can all be functions of t. In general the solution of such an equation is likely to be difficult to obtain, and we shall not undertake the discussion of any special cases here.

However, when we have a population process which is essentially multiplicative except for the inclusion of an immigration element, the following argument may be useful. Suppose that the probability-generating function at time t for the number of descendants of one individual at time s ($s \leqslant t$) is $Q(x, t; s)$. Then for an individual introduced, or existing, at time $s = 0$, we have

$$Q(x, t; 0) = P(x, t), \tag{9.46}$$

where $P(x, t)$ is the probability-generating function previously obtained in (9.31) with $a = 1$ for the non-homogeneous birth-and-death process. That is,

$$P(x, t) = 1 + \cfrac{1}{\cfrac{e^{\rho(t)}}{x - 1} - \int_0^t \lambda(\tau)e^{\rho(\tau)} \, d\tau}. \tag{9.47}$$

More generally, we have, by a slight extension of the argument leading to equation (9.31),

$$Q(x, t; s) = 1 + \cfrac{1}{\cfrac{e^{\rho(s,t)}}{x - 1} - \int_s^t \lambda(\tau)e^{\rho(s,t)} \, d\tau}, \tag{9.48}$$

where

$$\rho(s, t) = \int_s^t \{\mu(\tau) - \lambda(\tau)\} \, d\tau. \tag{9.49}$$

The main difference between this derivation and the previous one is that, while λ and μ are still functions of t, we assume a single individual at time $t = s$.

Now consider the interval $(\tau_r, \tau_r + \Delta\tau)$. The chance of an immigrant being introduced is $v(\tau_r)\,\Delta\tau$, and the probability-generating function for the descendants of such an individual at time t is $Q(x, t; \tau_r)$. The chance of no immigration is $1 - v(\tau_r)\,\Delta\tau$. Hence the net result is a probability-generating function at time t, *relating to the interval* $(\tau_r, \tau_r + \Delta\tau)$, given by

$$v(\tau_r)\,\Delta\tau\, Q(x, t; \tau_r) + \{1 - v(\tau_r)\,\Delta\tau\} = 1 + v(\tau_r)\,\Delta\tau\{Q(x, t; \tau_r) - 1\}.$$
$$(9.50)$$

We can now combine the contribution from the initial group of particles, say c, at $t = 0$, whose probability-generating function is

$$\{Q(x, t; 0)\}^c = \{P(x, t)\}^c,$$

with the contributions due to immigration over the whole range $(0, t)$. This gives a probability-generating function $P_e(x, t)$, which may be expressed as

$$P_e(x, t) = \{P(x, t)\}^c \lim_{\Delta\tau \to 0} \prod_r [1 + v(\tau_r)\Delta\tau\{Q(x, t; \tau_r) - 1\}]$$

$$= \{P(x, t)\}^c \exp \int_0^t v(\tau)\{Q(x, t; \tau) - 1\}\, d\tau. \qquad (9.51)$$

For given functions $\lambda(t)$, $\mu(t)$, and $v(t)$, we can calculate in succession $\rho(t)$, $P(x, t)$, $\rho(s, t)$, $Q(x, t; s)$, and $P_e(x, t)$. This may be difficult in specific cases, but at least we have obtained a solution in closed form which is susceptible of further development or computation.

PROBLEMS FOR SOLUTION

1. Investigate the properties of the non-homogeneous birth-and-death process described in Section 9.3 when the birth- and death-rates have the special linear forms $\lambda(t) = a + bt$, $\mu(t) = c + dt$.

What restrictions, if any, must be placed on the constants a, b, c, and d?

2. Write down the appropriate partial differential equation for the cumulant-generating function of the non-homogeneous birth-and-death process of Section 9.4 which includes immigration.

Obtain expressions for the mean and variance of population size in terms of $\lambda(t)$, $\mu(t)$, and $v(t)$.

3. What happens in Problem 2 if the immigration rate $v(t)$ oscillates, and is given by $v(t) = \alpha + \beta \sin wt$, while $\lambda(t) \equiv \lambda$ and $\mu(t) \equiv \mu$?

CHAPTER 10

Multi-Dimensional Processes

10.1 Introduction

So far we have confined the discussions of Markov processes in continuous time to the treatment of a single random variable $X(t)$, except for a brief indication at the end of Section 7.4 that the partial differential equation in (7.28) could be readily generalized to the case of two variables as shown in (7.31). There is in fact usually little difficulty in writing down the appropriate differential-difference equations for the state probabilities, or partial differential equations for the various kinds of generating functions. The main problem is to devise methods of solution. In the present chapter we shall examine a few relatively simple applications, and shall see how in some instances, e.g. the bacterial mutation problem treated in Section 10.4, certain kinds of approximations may be utilized.

By way of preliminaries, we first note that if there are two random variables $X(t)$ and $Y(t)$ we shall be concerned with the joint probability distribution at time t given by

$$P\{X(t) = m, Y(t) = n\} = p_{mn}(t),$$
(10.1)

for which we can define the probability-generating function

$$P(x, y, t) = \sum_{m,n} p_{mn}(t)x^m y^n.$$
(10.2)

Writing $x = e^\theta$ and $y = e^\phi$ gives the corresponding moment-generating function

$$M(\theta, \phi, t) = P(e^\theta, e^\phi, t) = \sum_{m,n} p_{mn}(t)e^{m\theta + n\phi},$$
(10.3)

while the cumulant-generating function is simply given by

$$K(\theta, \phi, t) = \log M(\theta, \phi, t).$$
(10.4)

Expanding $M(\theta, \phi, t)$ or $K(\theta, \phi, t)$ in powers of θ and ϕ provides the well-known definitions in terms of joint moments or cumulants, i.e.

$$M(\theta, \phi, t) = \sum_{u,v \geqslant 0}^{\infty} \mu'_{uv}(t)\theta^u\phi^v/u!\,v!, \qquad (10.5)$$

and

$$K(\theta, \phi, t) = \sum_{u,v \geqslant 0}^{\infty} \kappa_{uv}(t)\theta^u\phi^v/u!\,v!, \qquad (10.6)$$

where $\mu'_{00} \equiv 1$, $\kappa_{00} \equiv 0$.

Next, let us suppose that the joint probability distribution of relevant transitions in the interval Δt is given by

$$P\{\Delta X(t) = j, \Delta Y(t) = k \,|\, X(t), Y(t)\} = f_{jk}(X, Y)\Delta t, \quad j, k \text{ not both zero.}$$
$$(10.7)$$

The forward differential-difference equations for the joint probabilities $p_{mn}(t)$ can be obtained in a manner entirely analogous to that used in Section 7.2 for the one-variable case, namely, by writing $p_{mn}(t + \Delta t)$ in terms of the probabilities at time t and the corresponding transition probabilities. Thus we can write down immediately

$$p_{mn}(t + \Delta t) = \sum_{j,k} p_{m-j,n-k}(t)f_{jk}(m - j, n - k)\,\Delta t$$
$$+ p_{mn}(t)\Big\{1 - \sum_{j,k} f_{jk}(m, n)\,\Delta t\Big\},$$

whence

$$\frac{dp_{mn}}{dt} = -p_{mn}\sum_{j,k} f_{jk}(m, n) + \sum_{j,k} p_{m-j,n-k}f_{jk}(m - j, n - k), \quad (10.8)$$

it being understood that the summations exclude the case $j = k = 0$.

Starting from the above notions we can use the random-variable technique of Section 7.4 to derive partial differential equations for the probability-generating function or moment-generating function. It is unnecessary to describe this in detail. The equation corresponding to (7.28) comes out to be

$$\frac{\partial M(\theta, \phi, t)}{\partial t} = \sum_{j,k} (e^{j\theta + k\phi} - 1)f_{jk}\left(\frac{\partial}{\partial \theta}, \frac{\partial}{\partial \phi}\right)M(\theta, \phi, t),$$

$$j, k \text{ not both zero,} \quad (10.9)$$

as already given in (7.31), and the analog of (7.29) is

$$\frac{\partial P(x, y, t)}{\partial t} = \sum_{j,k} (x^j y^k - 1) f_{jk}\left(x \frac{\partial}{\partial x}, y \frac{\partial}{\partial y}\right) P(x, y, t),$$

$$j, k \text{ not both zero.} \quad (10.10)$$

We shall now apply those general results to a number of special cases.

10.2 Population growth with two sexes

Let us consider the growth of a population involving two kinds of individuals, namely, females and males, the corresponding random variables being $X(t)$ and $Y(t)$. We shall adopt, as an approximation, the following two-dimensional birth-and-death process (see Bharucha–Reid, 1960, Section 4.2, for further discussion and references). Let the birth-rate, for females only of course, be λ; and let the sex-ratio of females to males be p to $q = 1 - p$ at birth. This means that the chance of a given female producing a female or male birth in Δt is $\lambda p \, \Delta t$ or $\lambda q \, \Delta t$, respectively. We next assume female and male death-rates of μ and μ', respectively.

Thus the chance in Δt of the whole population producing a female is $\lambda p X \, \Delta t$, and a male is $\lambda q X \, \Delta t$, while the chance of a female dying is $\mu X \, \Delta t$, and of a male dying is $\mu' Y \, \Delta t$. The function $f_{jk}(X, Y)$ defined in (10.7) then takes only the following non-zero values:

$$f_{1,0} = \lambda p X, \qquad f_{-1,0} = \mu X, \qquad f_{0,1} = \lambda q X, \qquad f_{0,-1} = \mu' Y. \quad (10.11)$$

We can now write down the equation for the moment-generating function immediately from (10.9) as

$$\frac{\partial M}{\partial t} = (e^\theta - 1)\lambda p \frac{\partial M}{\partial \theta} + (e^{-\theta} - 1)\mu \frac{\partial M}{\partial \theta}$$

$$+ (e^\phi - 1)\lambda q \frac{\partial M}{\partial \theta} + (e^{-\phi} - 1)\mu' \frac{\partial M}{\partial \phi}.$$

Rearranging the terms, and writing $K = \log M$, gives the more convenient expression for the cumulant-generating function as

$$\frac{\partial K}{\partial t} = (\lambda p e^\theta + \lambda q e^\phi + \mu e^{-\theta} - \lambda - \mu)\frac{\partial K}{\partial \theta} + \mu'(e^{-\phi} - 1)\frac{\partial K}{\partial \phi}, \quad (10.12)$$

with initial condition

$$K(\theta, \phi, 0) = a\theta + b\phi, \quad (10.13)$$

if there are a females and b males at time $t = 0$.

We now equate coefficients of θ, ϕ, θ^2, $\theta\phi$ and ϕ^2 on both sides of (10.12) to give the differential equations

$$\frac{d\kappa_{10}}{dt} = (\lambda p - \mu)\kappa_{10}$$

$$\frac{d\kappa_{01}}{dt} = \lambda q \kappa_{10} - \mu'\kappa_{01}$$

$$\frac{d\kappa_{20}}{dt} = (\lambda p + \mu)\kappa_{10} + 2(\lambda p - \mu)\kappa_{20} \qquad (10.14)$$

$$\frac{d\kappa_{11}}{dt} = (\lambda p - \mu - \mu')\kappa_{11} + \lambda q \kappa_{20}$$

$$\frac{d\kappa_{02}}{dt} = \lambda q \kappa_{10} + \mu'\kappa_{01} + 2\lambda q \kappa_{11} - 2\mu'\kappa_{02}$$

These equations can be solved exactly, using the initial conditions that when $t = 0$, $\kappa_{10} = a$, and $\kappa_{01} = b$, all other cumulants being zero. The first two equations may be solved successively to give

$$\kappa_{10} = ae^{(\lambda p - \mu)t}, \qquad (10.15)$$

and
$$\kappa_{01} = be^{-\mu't} + \frac{\lambda qa}{\lambda p - \mu + \mu'}\{e^{(\lambda p - \mu)t} - e^{-\mu't}\}. \qquad (10.16)$$

The other second-order cumulants may also be calculated in a straightforward way, though the expressions are a little cumbersome, and we shall not reproduce them here.

It follows from (10.15) and (10.16) that the ratio of the numbers of males to females at time t is given by the quantity $R(t)$, where

$$R(t) = \frac{b}{a} e^{-(\lambda p - \mu + \mu')t} + \frac{\lambda q}{\lambda p - \mu + \mu'}\{1 - e^{-(\lambda p - \mu + \mu')t}\}. \qquad (10.17)$$

If $\lambda p - \mu + \mu' > 0$, this ratio has a finite limit as $t \to \infty$, namely

$$\lim_{t \to \infty} R(t) = \frac{\lambda q}{\lambda p - \mu + \mu'}. \qquad (10.18)$$

And if the male and female death rates are equal, i.e. $\mu' = \mu$, this limit is just q/p, the ratio of the relative birth-rates of male and female children.

10.3 The cumulative population

In some biological problems, where we are handling a stochastic model of population growth, we are interested not only in the size of the population at any instant of time but also in the total number of individuals who have ever lived up to that time. Thus, in dealing with bacterial growth, we may employ a method of measuring the population size, based on growing fresh cultures from samples, for instance, that provides us with an estimate of the total number of bacteria alive at the time of investigation: the so-called viable count. On the other hand, some optical methods yield estimates which include dead organisms as well (provided these have not disintegrated). The latter is the total count, and it is of some interest to study the distribution of this quantity as well as the actual population living at any instant. Again, in epidemic theory, it is often easier to observe with accuracy the total size of the epidemic, i.e. the total number of individuals who have ever been infected after the elapse of a sufficiently long period of time, than it is to record the actual numbers of infectious persons circulating in the population at any moment.

The investigation of the total count, at least for Markov processes, is by no means as difficult an undertaking as might be supposed. The basic device is simply to introduce a random variable, say $Y(t)$, to represent the total count, while $X(t)$ is the size of the actual population alive at time t. We then consider the two-dimensional process involving both $X(t)$ and $Y(t)$. Since the purpose of $Y(t)$ is to record the total count, it is clear that every time there is a birth, i.e. a transition in which $X(t)$ increases by a unit, $Y(t)$ also increases by a unit. But when $X(t)$ is decreased by a unit, as when a death occurs, $Y(t)$ remains unchanged. Thus if the main process for $X(t)$ is specified, we automatically have the specification for the joint behavior of $X(t)$ and $Y(t)$. See Kendall (1948b).

Let us suppose, for example, that we have a population which is sufficiently well described by a birth-and-death process (not necessarily homogeneous). Then following the procedure already employed in the last section, the only non-zero f_{jk} functions are

$$f_{1,1} = \lambda X, \qquad f_{-1,0} = \mu X. \tag{10.19}$$

The basic partial differential equation for the moment-generating function is thus, using (10.9),

$$\frac{\partial M}{\partial t} = \{\lambda(e^{\theta+\phi} - 1) + \mu(e^{-\theta} - 1)\} \frac{\partial M}{\partial \theta}, \tag{10.20}$$

and the corresponding equation for the cumulant-generating function is

$$\frac{\partial K}{\partial t} = \{\lambda(e^{\theta+\phi} - 1) + \mu(e^{-\theta} - 1)\} \frac{\partial K}{\partial \theta}, \tag{10.21}$$

with initial condition

$$K(\theta, \phi, 0) = a\theta + a\phi, \tag{10.22}$$

since $X(0) = a = Y(0)$.

This equation appears to be somewhat intractable in general, when λ and μ are both functions of t. Nevertheless, we can obtain a certain amount of useful information from the usual technique of equating coefficients on both sides of (10.21) to derive differential equations for the cumulants. Taking this procedure as far as the second-order cumulants only, we find the following equations

$$
\left.
\begin{aligned}
\frac{d\kappa_{10}}{dt} &= (\lambda - \mu)\kappa_{10} \\[2mm]
\frac{d\kappa_{01}}{dt} &= \lambda\kappa_{10} \\[2mm]
\frac{d\kappa_{20}}{dt} &= (\lambda + \mu)\kappa_{10} + 2(\lambda - \mu)\kappa_{20} \\[2mm]
\frac{d\kappa_{11}}{dt} &= \lambda\kappa_{10} + (\lambda - \mu)\kappa_{11} + \lambda\kappa_{20} \\[2mm]
\frac{d\kappa_{02}}{dt} &= \lambda\kappa_{10} + 2\lambda\kappa_{11}
\end{aligned}
\right\}
\qquad (10.23)
$$

We have already obtained solutions to the first and third equations in considering the non-homogeneous birth-and-death process in Section 9.3. In fact, replacing κ_1 and κ_2 in (9.37) and (9.38) by κ_{10} and κ_{20}, we can use the solutions in (9.40) and (9.41) to give

$$
\left.
\begin{aligned}
\kappa_{10} &= ae^{-\rho(t)} \\[2mm]
\kappa_{20} &= ae^{-2\rho(t)} \int_0^t \{\lambda(\tau) + \mu(\tau)\}e^{\rho(\tau)}\, d\tau
\end{aligned}
\right.
$$

where, as before,

$$
\left.
\rho(t) = \int_0^t \{\mu(\tau) - \lambda(\tau)\}\, d\tau
\right\}
$$

$$(10.24)$$

The second equation of (10.23) can now be solved in terms of known functions as

$$
\kappa_{01} = a + \int_0^t \lambda(\tau)\kappa_{10}\, d\tau, \qquad (10.25)
$$

while the covariance κ_{11} can be obtained from the fourth equation of (10.23) in the terms

$$
\kappa_{11} = e^{-\rho(t)} \int_0^t \lambda(\tau)e^{\rho(\tau)}(\kappa_{10} + \kappa_{20})\, d\tau. \qquad (10.26)
$$

And finally from the fifth equation we can derive an expression for κ_{02}, namely

$$\kappa_{02} = \int_0^t \lambda(\tau)(\kappa_{10} + 2\kappa_{11}) \, d\tau. \tag{10.27}$$

The homogeneous case

In the homogeneous case we have $\lambda(t) \equiv \lambda$ and $\mu(t) \equiv \mu$. Substituting in the general expressions (10.24)–(10.27), we obtain the specific results

$$\kappa_{10} = ae^{(\lambda-\mu)t}, \qquad \kappa_{20} = \frac{a(\lambda+\mu)}{\lambda-\mu} e^{(\lambda-\mu)t}(e^{(\lambda-\mu)t} - 1), \tag{10.28}$$

as previously given in (8.48) and (8.49), together with the new expressions

$$\left. \begin{aligned} \kappa_{01} &= \frac{a}{\mu-\lambda}(\mu - \lambda e^{(\lambda-\mu)t}) \\[2mm] \kappa_{11} &= \frac{a\lambda e^{(\lambda-\mu)t}}{\mu-\lambda}\left\{2\mu t - \frac{\mu+\lambda}{\mu-\lambda}(1 - e^{(\lambda-\mu)t})\right\} \\[2mm] \kappa_{02} &= \frac{a\lambda(\mu+\lambda)}{(\mu-\lambda)^2}(1 - e^{(\lambda-\mu)t}) - \frac{4a\lambda^2\mu t e^{(\lambda-\mu)t}}{(\mu-\lambda)^2} \\[2mm] &\qquad + \frac{a\lambda^2(\mu+\lambda)}{(\mu-\lambda)^3}(1 - e^{2(\lambda-\mu)t}) \end{aligned} \right\}. \tag{10.29}$$

It is interesting to observe that as $t \to \infty$, we have for $\lambda < \mu$,

$$\kappa_{01} \to \frac{a\mu}{\mu-\lambda}, \qquad \kappa_{11} \to 0, \qquad \kappa_{02} \to \frac{a\mu\lambda(\mu+\lambda)}{(\mu-\lambda)^3}, \tag{10.30}$$

so that the average number of individuals who ever exist in the whole process (for which extinction is certain if $\lambda < \mu$) is $a\mu/(\mu - \lambda)$, including the a individuals present at $t = 0$.

Actually, in this homogeneous case we can find an explicit solution for the distribution at time t. Suppose we work directly with the probability-generating function $P(x, y)$. From (10.20), with $e^\theta = x$ and $e^\phi = y$, we obtain

$$\frac{\partial P}{\partial t} = \{\lambda(x^2 y - x) + \mu(1 - x)\} \frac{\partial P}{\partial x}, \tag{10.31}$$

with initial condition

$$P(x, y, 0) = x^a y^a. \tag{10.32}$$

The subsidiary equations are

$$\frac{dt}{1} = \frac{-dx}{\lambda y x^2 - (\lambda + \mu)x + \mu} = \frac{dy}{0} = \frac{dP}{0}. \tag{10.33}$$

Two independent integrals are clearly

$$y = \text{constant}, \qquad P = \text{constant}, \tag{10.34}$$

and a third can be obtained from

$$dt = \frac{dx}{-\lambda y x^2 + (\lambda + \mu)x - \mu}$$

$$\equiv \frac{dx}{\lambda y(x - \alpha)(\beta - x)}, \tag{10.35}$$

where $0 < \alpha < 1 < \beta$, and α, β are the roots of the quadratic

$$\lambda y x^2 - (\lambda + \mu)x + \mu = 0. \tag{10.36}$$

Integrating (10.35) gives

$$t = \frac{1}{\lambda y(\beta - \alpha)} \log \frac{x - \alpha}{\beta - x} + \text{constant},$$

or

$$\frac{x - \alpha}{\beta - x} e^{-\lambda y(\beta - \alpha)t} = \text{constant}. \tag{10.37}$$

The general solution can therefore be written as

$$P(x, y, t) = \Psi\left(\frac{x - \alpha}{\beta - x} e^{-\lambda y(\beta - \alpha)t}, y\right). \tag{10.38}$$

Using the initial condition (10.32) gives

$$P(x, y, 0) \equiv x^a y^a = \Psi\left(\frac{x - \alpha}{\beta - x}, y\right). \tag{10.39}$$

We now put

$$\frac{x - \alpha}{\beta - x} = w, \qquad x = \frac{\alpha + \beta w}{1 + w}, \tag{10.40}$$

and substitute in (10.39) to give the explicit functional form

$$\Psi(w, y) = \left(\frac{\alpha + \beta w}{1 + w}\right)^a y^a. \tag{10.41}$$

Using this on the right of (10.38) finally yields

$$P(x, y, t) = y^a \left\{ \frac{\alpha(\beta - x) + \beta(x - \alpha)e^{-\lambda y(\beta - \alpha)t}}{(\beta - x) + (x - \alpha)e^{-\lambda y(\beta - \alpha)t}} \right\}^a, \qquad (10.42)$$

where α and β are as defined above, i.e. the roots of (10.36).

This time we can obtain the whole asymptotic distribution in the case of certain extinction, $\lambda < \mu$, as $t \to \infty$. So far as the marginal distribution of the cumulative population is concerned, we simply require the probability-generating function

$$P(1, y, \infty) = (y\alpha)^a, \qquad (10.43)$$

where
$$\alpha = \frac{(\lambda + \mu) - \{(\lambda + \mu)^2 - 4\lambda\mu y\}^{\frac{1}{2}}}{2\lambda y}.$$

To expand (10.43) in powers of y is easy for $a = 1$, though a little more difficult if $a > 1$. In the latter, general, case the required result is most easily derived by using a suitable form of Lagrange's expansion (see equation (4.40)). We want to expand $\psi(\alpha) = \alpha^a$, where

$$y = \frac{\alpha - \mu/(\lambda + \mu)}{\alpha^2 \lambda/(\lambda + \mu)} \equiv \frac{\alpha - \alpha_0}{\phi(\alpha)}.$$

Direct application of the expansion in (4.40) now gives the coefficient q_r of y^r in $y^a \alpha^a$ in (10.43) as

$$q_r = \frac{a(2r - a - 1)! \, \lambda^{r-a} \mu^r}{(r - a)! \, r! \, (\lambda + \mu)^{2r-a}}, \quad a \leqslant r. \qquad (10.44)$$

which is the desired probability distribution. (For the special case of $a = 1$, see Kendall, 1948b.)

10.4 Mutation in bacteria

One important and interesting application of a two-dimensional stochastic model is to the occurrence and subsequent spread of mutant genes or mutant types of organism in a population subject to some kind of birth-and-death process. In a very general treatment we should take into account a variety of factors, including selection pressures (we have already examined one small aspect of this problem in Section 6.3, where we considered the chance of mutant types becoming extinct). However, for the purpose of the present section, we shall look at a much more limited application to a specific problem, namely the occurrence and spread of mutant types in a bacterial colony, in which all individuals are subject to growth by a simple birth process, and there are no selection effects. This we can do without any elaborate discussion of the genetical principles involved.

We suppose that there are essentially just two kinds of bacteria: the normal kind, and a mutant variety which appears rather rarely as a chance random transformation of a normal organism. Now there are various models that we might adopt. (See Armitage, 1951, for an extensive discussion of the whole area.) For instance, we could develop a model along the lines of the two-dimensional process in Section 10.2. There would be two populations to be considered, one of normals and one of mutants, each subject to a birth-and-death process, with the additional feature of rare random transitions in which a normal turned into a mutant. There are, however, certain difficulties in a full treatment of this model (see Bartlett, 1955, Section 4.31). Now, since mutations are rare, they will only occur with appreciable probability in relatively large groups of normal bacteria. In such large groups statistical fluctuations in group size about the mean value will be relatively small. This suggests that we might introduce a useful simplification by assuming that the normal population grows purely deterministically. We should then have reduced a two-dimensional stochastic problem to a one-dimensional stochastic problem. Moreover, it is not unreasonable, at least as a first approximation to assume that while both normals and mutants are subject to birth with the same rate λ, there is no death involved. Thus if we start at $t = 0$ with a normals, the size of the normal population at time t is $ae^{\lambda t}$.

The mutant population must be treated stochastically, since it is in general not very large, and we suppose that the chance of any mutant dividing into two new mutant individuals in time Δt is $\lambda \, \Delta t$. We next consider the chance appearance of mutants in the normal population. Suppose that when a normal bacterium divides it usually produces two normals, but with a small probability p there is one normal and one mutant. Now the average number of births in time Δt in a normal population of size $ae^{\lambda t}$ is $\lambda ae^{\lambda t} \, \Delta t$. It therefore follows that the chance of a mutant individual appearing in the normal population in Δt is $p\lambda ae^{\lambda t} \, \Delta t$. Strictly speaking, this formulation means that the normal population size at time t will be slightly less than $ae^{\lambda t}$, but for small p we can ignore this complication.

In formulating the basic partial differential equation for the moment-generating function of the size of the mutant population at any instant, there is only one type of transition to be considered, namely $j = +1$, for which we have

$$f_1 = \lambda X + \lambda pae^{\lambda t}. \tag{10.45}$$

The required equation is thus

$$\frac{\partial M}{\partial t} = \lambda pae^{\lambda t}(e^{\theta} - 1)M + \lambda(e^{\theta} - 1)\frac{\partial M}{\partial \theta}, \tag{10.46}$$

with initial condition

$$M(\theta, 0) = 1, \tag{10.47}$$

since we can assume that we start off with no mutants.

Putting $K = \log M$ in (10.46), we have the equation for the cumulant-generating function given by

$$\frac{\partial K}{\partial t} = \lambda p a e^{\lambda t}(e^{\theta} - 1) + \lambda(e^{\theta} - 1)\frac{\partial K}{\partial \theta}, \tag{10.48}$$

with initial condition

$$K(\theta, 0) = 0. \tag{10.49}$$

The subsidiary equations are

$$\frac{dt}{1} = \frac{-d\theta}{\lambda(e^{\theta} - 1)} = \frac{dK}{\lambda p a e^{\lambda t}(e^{\theta} - 1)}. \tag{10.50}$$

The first two expressions in (10.50) give

$$\lambda \, dt = \frac{-d\theta e^{-\theta}}{1 - e^{-\theta}}, \tag{10.51}$$

which integrates to

$$\lambda t = -\log(1 - e^{-\theta}) + \text{constant},$$

or $\qquad (1 - e^{-\theta})e^{\lambda t} = \text{constant} \equiv A, \tag{10.52}$

where it is convenient to specify the constant by A. The second and third expressions in (10.50) lead to

$$\frac{dK}{d\theta} = -pae^{\lambda t},$$

which involves t as well as K and θ. In applying the present method of integration, we cannot treat t as constant in this equation but can eliminate it by means of the relation in (10.52). We then have

$$\frac{dK}{d\theta} = \frac{-Apa}{1 - e^{-\theta}} = \frac{-Apae^{\theta}}{e^{\theta} - 1}, \tag{10.53}$$

which integrates to yield

$$K = -Apa \log(e^{\theta} - 1) + \text{constant}$$

$$= -pae^{\lambda t}(1 - e^{-\theta})\log(e^{\theta} - 1) + \text{constant},$$

using (10.52) again to eliminate A, or

$$K + pae^{\lambda t}(1 - e^{-\theta})\log(e^{\theta} - 1) = \text{constant}. \tag{10.54}$$

From the two integrals in (10.52) and (10.54), we can write the required general solution as

$$K + pae^{\lambda t}(1 - e^{-\theta})\log(e^{\theta} - 1) = \Psi\{(1 - e^{-\theta})e^{\lambda t}\}. \tag{10.55}$$

The initial condition (10.49) applied to (10.55) gives

$$\Psi(1 - e^{-\theta}) = pa(1 - e^{-\theta})\log(e^{\theta} - 1). \tag{10.56}$$

We next put

$$1 - e^{-\theta} = u, \qquad e^{\theta} - 1 = \frac{u}{1 - u}, \tag{10.57}$$

in (10.56) to reveal the function Ψ as

$$\Psi(u) = pau \log \frac{u}{1 - u}. \tag{10.58}$$

We now return to (10.55), and, using (10.58), can finally express K in the form

$$K(\theta, t) = pae^{\lambda t}(e^{-\theta} - 1)\log(e^{\theta} - 1) + pa(1 - e^{-\theta})e^{\lambda t} \log \frac{(1 - e^{-\theta})e^{\lambda t}}{1 - (1 - e^{-\theta})e^{\lambda t}}$$

$$= pae^{\lambda t}(e^{-\theta} - 1)\log\{1 + e^{\theta}(e^{-\lambda t} - 1)\}. \tag{10.59}$$

Writing $e^{\theta} = x$ in (10.59), and taking exponentials of both sides, gives the probability-generating function as

$$P(x, t) = \{1 + (e^{-\lambda t} - 1)x\}^{pae^{\lambda t}(1-x)/x}. \tag{10.60}$$

The probability-generating function is of a somewhat complicated form for obtaining the individual probabilities, though we can effect a Taylor expansion if we first write it in the form

$$P(x, t) = \exp\left[pae^{\lambda t} \frac{1 - x}{x} \log\{1 + (e^{-\lambda t} - 1)x\}\right]. \tag{10.61}$$

The chance of no mutants appearing, $p_0(t) = P(0, t)$, is easily found to be

$$p_0(t) = e^{-pa(e^{\lambda t} - 1)}. \tag{10.62}$$

Cumulants, of course, are easily obtained from (10.59). The first two are

$$m(t) = \lambda pate^{\lambda t}, \tag{10.63}$$

and

$$\sigma^2(t) = 2pae^{\lambda t}(e^{\lambda t} - 1) - \lambda pate^{\lambda t}. \tag{10.64}$$

If we write $N = ae^{\lambda t}$, where N is the total number of normal bacteria, we can express the mean and variance of the mutant population, for large t, as

$$m(t) = pN \log(N/a) \atop \sigma^2(t) \sim 2pN^2/a \Bigg\} . \qquad (10.65)$$

It is also of interest to note that, as t becomes large, the probability-generating function in (10.61) assumes the asymptotic form

$$P_N(x) \sim \exp\left\{\frac{pN(1-x)}{x} \log(1-x)\right\}. \qquad (10.66)$$

This distribution is easily seen to have no finite moments, although it is a true probability distribution. For example, $\lim_{x \to 1} P_N(x) = 1$.

The asymptotic distribution given by (10.66) has been tabulated by Lea and Coulson (1949) for different values of $M = pN$. They showed that, for $M \geqslant 4$, a reasonably good approximation was to take

$$w(M) = \frac{11 \cdot 6}{(r/M) - \log M + 4 \cdot 5} - 2 \cdot 02, \qquad (10.67)$$

where r is the observed number of mutants in a colony, as a normal variable with zero mean and unit standard deviation. This provides a convenient method of estimating the mutation-rate p from parallel cultures (assuming that we can also make a sufficiently good estimate of N).

An alternative approximation

Now the use of a simple birth process in dealing with the reproduction of the mutant organisms above assumes that the distribution of the life-time from birth to division into two daughter cells is negative exponential (see Section 8.2). There are reasons, elaborated in the next section, for supposing that this assumption is unrealistic, and that, although the life-times show some variation, a better first approximation would be to regard both normals and mutants as reproducing approximately deterministically. A model involving purely deterministic growth would still retain a stochastic element due to the random occurrence of mutations. It is worth investigating the consequences of such a set-up, which looks as though it might offer a further useful simplification. In fact, there are one or two awkward, but interesting, special features.

We suppose, therefore, that growth takes place deterministically according to the exponential law for both normals and mutants. As before, we write $N = ae^{\lambda t}$ for the total number of normal bacteria at time t. Let the number of mutants at time t be represented by the random variable $X(t)$. Then, in so far as the mutants themselves reproduce deterministically according to the exponential law, the variable $X(t)$ will be continuous rather than discrete. But when a mutation occurs $X(t)$ will jump from X to $X + 1$. The distribution of the number of mutants thus involves both continuous and discrete elements. From the point of view of mathematical rigor there are certain analytical difficulties here. We shall ignore these, and adopt the following somewhat heuristic treatment.

The variable $X(t)$ is subject to two possible kinds of transition in the interval Δt. First, there may be a jump of $+1$ due to a random mutation, occurring with probability $p\lambda N \, \Delta t$ (this is of course the same value as that used in the previous model). Secondly, a deterministic increment of $Xe^{\lambda \Delta t} - X = \lambda X \, \Delta t$ occurs with probability one. We therefore have an overall increment of $1 + \lambda X \, \Delta t$ with probability $p\lambda N \, \Delta t$, and $\lambda X \, \Delta t$ with probability $1 - p\lambda N \, \Delta t$. We cannot use the standard form of the partial differential equation for the moment-generating function appearing in (7.28), since this applies only to discrete variables. Returning, however, to the more general formulas in (7.25) and (7.26), the basic quantity required is

$$\Psi(\theta, t, X) = \lim_{\Delta t \to 0} \; E_{\Delta t \mid t} \left\{ \frac{e^{\theta \, \Delta X(t)} - 1}{\Delta t} \right\}$$

$$= \lim_{\Delta t \to 0} \frac{p\lambda N \Delta t e^{(1 + \lambda X \Delta t)\theta} + (1 - p\lambda N \Delta t)e^{\lambda X \Delta t \theta} - 1}{\Delta t}$$

$$= p\lambda N(e^{\theta} - 1) + \lambda \theta X. \tag{10.68}$$

We then have immediately, from (7.26),

$$\frac{\partial M}{\partial t} = \lambda p a e^{\lambda t}(e^{\theta} - 1)M + \lambda \theta \frac{\partial M}{\partial \theta}, \tag{10.69}$$

which differs from the equation for the previous model, i.e. (10.46), in respect of the last term on the right. The equation for the cumulant-generating function is

$$\frac{\partial K}{\partial t} = \lambda p a e^{\lambda t}(e^{\theta} - 1) + \lambda \theta \frac{\partial K}{\partial \theta}, \tag{10.70}$$

with the usual initial condition

$$K(\theta, 0) = 0. \tag{10.71}$$

There is no difficulty in solving (10.70) to find an expansion for K in powers of θ, which gives explicit formulas for the cumulants. The attempt to find a convenient closed expression for K, from which the distribution function might be derived is more troublesome. So far as the cumulants are concerned, we can equate coefficients of powers of θ on both sides of (10.70) to give

$$\left.\begin{array}{l} \dfrac{d\kappa_1}{dt} = \lambda pae^{\lambda t} + \lambda\kappa_1 \\[3mm] \dfrac{d\kappa_r}{dt} = \lambda pae^{\lambda t} + r\lambda\kappa_r, \quad r > 1 \end{array}\right\}, \qquad (10.72)$$

from which we readily obtain

$$\left.\begin{array}{l} \kappa_1 = \lambda pate^{\lambda t} \\[3mm] \kappa_r = \dfrac{pa}{r-1}(e^{r\lambda t} - e^{\lambda t}), \quad r > 1 \end{array}\right\}. \qquad (10.73)$$

The mean is, as we might expect, exactly the same as in the previous model (see (10.63)), though the variance is different.

10.5 A multiple-phase birth process

A shortcoming of all the continuous time birth-and-death processes considered so far has been that the assumption of random transitions entails a negative exponential distribution of the time between successive transitions. In particular, the stochastic birth process adopted for bacterial growth in the last section implied a negative exponential distribution of individual life-times. The only alternative to this was to work with a purely deterministic growth process.

While these two extremes may be sufficiently accurate for many purposes, such as gaining general insight into whatever process we are studying, we may in some circumstances, such as attempting to fit a model to actual data, require less restrictive assumptions.

First, let us consider what happens in practice. Kelly and Rahn (1932) have investigated the division and growth of *Bacterium aerogenes*, following the actual progress of each individual. A fair amount of variation about a mean generation time of approximately 20 minutes was observed, and the frequency distributions appeared to be unimodal. Neither a negative exponential distribution of generation-time nor a fixed value could be regarded as a sufficiently accurate description of the data. It was suggested by D. G. Kendall (1948a) that a suitably chosen χ^2-distribution would be more appropriate, and an examination of Kelly and Rahn's data showed that the number of degrees of freedom for different groups of data varied between about 25 and 100, the overall average being somewhere in the neighborhood of 50.

A general treatment of χ^2-generation times would have the advantage that the two cases already considered correspond to two degrees of freedom (negative exponential distribution) and the limiting situation of an infinite number of degrees of freedom (fixed generation-times).

Now the comparative tractability of the birth-and-death processes discussed so far has depended in large measure on their possessing the basic Markov property. The latter would be lost if we adopted a general distribution of life-times, though more advanced methods might be used, such as the integral equation approach described in Chapter 16. Suppose, however, that we adopt the following approach.

We imagine, following Kendall (1948a), that the life-time of each individual consists of k phases. A new individual starts life in Phase 1 and passes successively through each phase in turn. On leaving Phase k, the individual divides into two and is replaced by two new individuals in Phase 1, and so on. Let us now assume that the transitions from phase to phase occur at random. Or, to be more specific, let the chance of an individual in any phase at t moving to the next phase in Δt be $k\lambda\,\Delta t$. The distribution of stay in any phase is thus the negative exponential

$$f(\tau) = k\lambda e^{-k\lambda\tau}, \quad 0 \leqslant \tau < \infty. \tag{10.74}$$

Apart from a scale factor, this is essentially a χ^2-distribution with two degrees of freedom. Consequently, the time spent between starting in Phase 1 and finally leaving Phase k has essentially a χ^2-distribution with $2k$ degrees of freedom.

The mean of the distribution in (10.74) is $(k\lambda)^{-1}$. Thus the average life-time is λ^{-1}. This is independent of k, and so we can vary the shape of the overall distribution by changing k without altering the mean.

So far we have introduced the device of imagining k phases simply as a way of deriving a χ^2-distribution. But there is another vitally important consequence. Namely, that if we consider the k-dimensional process for the numbers of individuals in each phase, the Markov property is retained in virtue of the random transitions from phase to phase. The phases themselves need not have any practical biological significance: they are introduced solely for the purpose of constructing a model that has certain important realistic features, as well as being mathematically tractable.

Let the number of individuals in the jth phase be represented by the random variable X_j. Then the probability of a transition occurring in Δt in which one of these individuals jumps to the $(j + 1)$th phase is $k\lambda X_j\,\Delta t$. When this happens X_j is diminished by one unit and X_{j+1} is increased by one unit, if $j < k$. If $j = k$, we have X_k losing one unit, and X_1 gaining *two* units.

By an obvious extension of (7.28) to the case of a k-dimensional process, we may write the partial differential equation of the joint moment-generating function of the k variables as

$$\frac{\partial M}{\partial t} = k\lambda \sum_{j=1}^{k-1} (e^{-\theta_j + \theta_{j+1}} - 1) \frac{\partial M}{\partial \theta_j} + k\lambda(e^{-\theta_k + 2\theta_1}) \frac{\partial M}{\partial \theta_k}, \quad (10.75)$$

the boundary condition for a single individual in Phase 1 at $t = 0$ being

$$M(\theta_1, \theta_2, \cdots, \theta_k; 0) = e^{\theta_1}. \quad (10.76)$$

The probability-generating function equation may clearly be written

$$\frac{\partial P}{\partial t} = k\lambda \sum_{j=1}^{k-1} (x_{j+1} - x_j) \frac{\partial P}{\partial x_j} + k\lambda(x_1^2 - x_k) \frac{\partial P}{\partial x_k}, \quad (10.77)$$

with initial condition

$$P(x_1, x_2, \cdots, x_k; 0) = x_1. \quad (10.78)$$

Equation (10.77) looks fairly encouraging, as it is of the standard linear type appearing in (7.36). The subsidiary equations are

$$\frac{dP}{0} = \frac{k\lambda \, dt}{1} = \frac{dx_1}{x_1 - x_2} = \cdots = \frac{dx_{k-1}}{x_{k-1} - x_k} = \frac{dx_k}{x_k - x_1^2}. \quad (10.79)$$

Some simplification results if we write

$$k\lambda t = T, \qquad y_i = e^{-T} x_i. \quad (10.80)$$

We then have subsidiary equations in the form

$$\left.\begin{array}{l} \dfrac{dy_i}{dT} = -y_{i+1}, \quad 0 \leqslant i \leqslant k-1 \\[2mm] \dfrac{dy_k}{dT} = -e^T y_1^2 \\[2mm] \end{array}\right\} . \quad (10.81)$$

and $\qquad dP = 0$

It was shown by Kendall that the required solution can finally be put in the form

$$P = e^{-k\lambda t} U(k\lambda t), \quad (10.82)$$

where $U(z)$ is determined by

$$\left(\frac{d}{dz}\right)^k U(z) = e^{-z}\{U(z)\}^2, \quad (10.83)$$

subject to the conditions

$$\left[\left(\frac{d}{dz} \right)^i U(z) \right]_{z=0} = x_{i+1}, \quad 0 \leqslant i \leqslant k-1. \tag{10.84}$$

Although this looks like a fairly straightforward reduction to a simple problem, equation (10.83) appears to be intractable for $k \geqslant 2$. Nevertheless, the compact analytic form might yield to further investigation, and might be used for instance as a basis for an approximation procedure.

However, (10.75) can be used to obtain differential equations for the moments. Equating coefficients of the θ_j only gives

$$\left. \begin{aligned} (k\lambda)^{-1} \frac{dm_1}{dt} &= -m_1 + 2m_k \\[2mm] (k\lambda)^{-1} \frac{dm_j}{dt} &= -m_j + m_{j-1}, \quad 2 \leqslant j \leqslant k \end{aligned} \right\}, \tag{10.85}$$

where m_j is the mean of X_j, the number of individuals in Phase j at time t. These equations are soluble, though complicated. It can be shown that

$$m_j = k^{-1} 2^{(1-j)/k} \sum_{r=0}^{k-1} \omega^{-(j-1)r} \exp\{(2^{1/k}\omega^r - 1)k\lambda t\}, \tag{10.86}$$

where ω is the primitive kth root of unity, i.e. $e^{2\pi i/k}$. The expected total population, i.e. $m(t) = \sum_{j=1}^{k} m_j(t)$, then comes out to be

$$m(t) = e^{-k\lambda t} \sum_{n=0}^{\infty} 2^n \left\{ \frac{(k\lambda t)^{nk}}{(nk)!} + \cdots + \frac{(k\lambda t)^{nk+k-1}}{(nk+k-1)!} \right\}. \tag{10.87}$$

The occurrence of Poisson terms in this expression suggests that it might be obtainable by more elementary considerations.

Now it can be seen that the dominant term in (10.86) is always the first one. So for large t we have

$$m_j \sim k^{-1} 2^{-(j-1)/k} \exp\{(2^{1/k} - 1)k\lambda t\}, \tag{10.88}$$

and

$$m(t) \sim \frac{\exp\{(2^{1/k} - 1)k\lambda t\}}{2k(1 - 2^{-1/k})}. \tag{10.89}$$

The asymptotic value of the variance comes out to be

$$\sigma^2(t) \sim \left\{ \frac{2}{\{1 + 2(2^{1/k} - 1)\}^k - 2} - 1 \right\} m^2(t). \tag{10.90}$$

For a more detailed discussion of this process, the paper by Kendall (1948a) should be consulted.

The complexity of the above results, whose derivation has only been sketched in outline, seems to derive from the attempt to find a complete solution for the multi-phase birth process at an arbitrary time t. Although this process does not have a stationary distribution for large t, it may happen that some more complicated process incorporating a multi-phase aspect does have a limiting distribution. The device of introducing the artefact of k phases with random transitions from phase to phase in order to mimic a χ^2-distribution of generation time with $2k$ degrees of freedom is in fact extremely valuable, and has been used for example in the theory of queueing to obtain the equilibrium distribution of queue length with a χ^2-distribution of service-time.

PROBLEMS FOR SOLUTION

1. Obtain the results for the mean and variance of the number of mutant bacteria given in (10.63) and (10.64) directly from the partial differential equation in (10.48) by equating coefficients of θ and θ^2, and solving (i.e. without obtaining an explicit solution for the cumulant-generating function).

2. Examine the model for mutation in bacteria in which both the mutant population and the normal population grow stochastically with birth-rate λ.

Write down the partial differential equation for the moment-generating function, and find the means, variances, and covariance of the mutant and normal populations.

CHAPTER 11

Queueing Processes

11.1 Introduction

A great many applications of stochastic processes in the area of biology and medicine are to the kind of birth-and-death situations discussed in the last four chapters, where we are thinking primarily in terms of the behavior of some biological population of individuals. But not all applications are of precisely this nature. This is especially true of many of the problems arising in the field of activity loosely defined as "operations research", i.e. research concerned with applying scientific methods to the problems facing executive and administrative authorities.

In medicine, for example, there is both the problem of developing medical science as fully as possible, and the problem of organizing and distributing known methods of medical care as widely and as efficiently as possible. The latter aspect involves such matters as research into the function and design of hospitals and other medical institutions so as to meet the requirements of both patients and staff in an optimal way. As with the analogous situations in industry, such problems of organization frequently entail some kind of waiting line or queue. And so the stochastic theory of queueing processes is a subject of considerable practical importance.

It is therefore desirable to include some treatment of queueing theory in a book primarily concerned with applications to biology and medicine because of the operations research aspects. Moreover, the discussion is very conveniently introduced at the present stage. Partly because it is easy to set up the appropriate models in terms of the concepts that have already been analyzed in great detail in the previous work on birth-and-death processes, and partly because there are one or two novel features which necessitate special mathematical treatment, and which provide additional tools for the study of stochastic processes in general.

There is now an immense literature on the theory of queues. The

136

present chapter can touch only on some of the main ideas. For a more extensive treatment of the theory the reader should consult Chapter 9 of Bharucha-Reid (1960), or the books by Saaty (1961) and Cox and Smith (1961). Specific applications to problems of the organization of medical care, particularly in the function and design of hospitals, have been discussed by Bailey (1954).

11.2 Equilibrium theory

Let us first consider how we may specify a fairly simple type of waiting-line situation in mathematical terms. There are usually three main aspects of the system which may be called (i) the *input process*, (ii) the *queue discipline*, and (iii) the *service mechanism*. Let us take these in turn.

The input process is simply a mathematical description of the way customers arrive. In many cases this can be taken to be a Poisson process with some suitable parameter. Thus in certain circumstances the initiation of telephone calls, the occurrence of accidents, the demand for hospital beds, etc., can be regarded as being approximately of this kind. We can then suppose that the chance of a new customer arriving in time Δt is $\lambda \, \Delta t$. But if we are dealing with an appointment system, say for hospital outpatients, we may want to consider the effect of regular arrivals with constant inter-arrival times. More generally, we must specify the precise distribution of the interval between successive arrivals.

The queue discipline refers to the way in which a waiting-line is formed, maintained and dealt with. The simplest arrangement is the "first come, first served" rule: new arrivals join the end of the queue, and the customers at the head of the queue are served in order of arrival. It is of course possible to modify this simple scheme to allow for some system of priority in service.

The service mechanism deals with the output end of the whole process. It is usually specified in sufficient detail once we know the number of servers and the distribution of service-time.

If arrivals occur at random, and if the service-time is negative exponential in distribution so that the departures of served customers are effectively at random, it is clear that the whole model will be described by a Markov process in continuous time, of the general type we have already dealt with. But the general queueing process in continuous time does not have a Markov character (see Section 16.3 for a more general formulation of a non-Markovian queue in terms of an integral equation). However, it is often possible to pick out a series of points in discrete time for which the behavior of an appropriately chosen variable is adequately described by a

Markov chain. This is the method of an *imbedded Markov chain*. In the case of a single-server queue with random arrival but a general distribution of service-time, it is easy to see that the queue length left behind by a departing customer can be used to specify a Markov chain, the properties of which can be studied by the theory and methods of Chapter 5.

In the present section we shall consider only certain features of the equilibrium distribution of queue length (assuming this distribution to exist) applicable where a long period of time has elapsed since the start of the process. Continuous-time treatments will be investigated later.

Let us suppose therefore that we have a single-server queue with some general distribution of service-time, and to begin with at any rate some general distribution of inter-arrival time. Let $X(t)$ be the number of customers in line, including the one being served, at time t. We now concentrate on the series of instants at which customers are about to depart. For any such epoch, τ say, we can write the length of queue left behind by the departing customer as

$$Q = X(\tau + \Delta\tau) = X(\tau) - 1. \qquad (11.1)$$

Now suppose that the service-time for the next customer is V, and that R new customers arrive during this time. Then if $Q \geqslant 1$, the value of Q for the queue left behind by the next departing customer is given by

$$Q' = Q + R - 1, \qquad Q \geqslant 1. \qquad (11.2)$$

But if $Q = 0$, we have

$$Q' = Q + R, \qquad Q = 0. \qquad (11.3)$$

Both results (11.2) and (11.3) can be conveniently combined by writing

$$Q' = Q + R - 1 + \delta(Q), \qquad (11.4)$$

where
$$\left.\begin{aligned}\delta(Q) &= 0, \quad Q \geqslant 1\\ &= 1, \quad Q = 0\end{aligned}\right\}. \qquad (11.5)$$

We can now take expectations in (11.4) to give

$$E(Q') = E(Q) + E(R) - 1 + E(\delta).$$

If equilibrium has been achieved we shall have $E(Q') = E(Q)$, so that the above expression can be written

$$\begin{aligned}E(\delta) &= 1 - E(R)\\ &= 1 - \rho, \quad \text{say,}\end{aligned} \qquad (11.6)$$

where ρ, the *traffic intensity*, is simply the mean number of arrivals during the mean service-time, $E(V)$.

Next, we square (11.4) and take the expectation, to give

$$E\{(Q')^2\} = E(Q^2) + E\{(1 - R)^2\} + E(\delta^2) + 2E(Q\delta) \\ + 2E\{(R - 1)\delta\} + 2E\{Q(R - 1)\}. \quad (11.7)$$

Now, from (11.5) we see that $\delta^2 = \delta$ and $Q\delta = 0$. Also, we can normally assume that R and Q are independent, and therefore that R and δ are independent. Further, in equilibrium we have $E\{(Q')^2\} = E(Q^2)$. Applying these results to (11.7) gives

$$0 = E\{(1 - R)^2\} + (1 - \rho) - 2(1 - \rho)^2 + 2E\{Q(\rho - 1)\},$$

which, on rearrangement, yields

$$E(Q) = \rho + \frac{E(R^2) - \rho}{2(1 - \rho)}. \quad (11.8)$$

Equation (11.8) is a fairly general result. To see its implications let us specialize the conditions a little. Suppose that the input process is in fact Poisson, with an average of λ arrivals in unit time. For given V the variable R therefore has a Poisson distribution with mean λV. It follows that

$$\rho = E(R) = \lambda E(V). \quad (11.9)$$

We also have

$$E(R^2) = \underset{V}{E}(\lambda V + \lambda^2 V^2),$$

since the second moment of R about the origin, given V, is $\lambda V + (\lambda V)^2$. Thus

$$E(R^2) = \rho + \lambda^2\{\sigma_V^2 + (\rho/\lambda)^2\}, \quad \text{using (11.9)}$$
$$= \rho + \rho^2 + \lambda^2\sigma_V^2, \quad (11.10)$$

where σ_V^2 is the variance of V. Substituting (11.10) in (11.8) and rearranging, yields the formula

$$E(Q) = \frac{\rho(2 - \rho) + \lambda^2\sigma_V^2}{2(1 - \rho)}. \quad (11.11)$$

Let us now consider two extreme cases (i) where the service-time is negative-exponentially distributed, and (ii) constant service-time. In the first case we have a negative exponential distribution with mean $\mu^{-1} = \rho/\lambda$ (using (11.9)). (Notice that this distribution implies that the chance of the customer being served actually departing in time Δt is just $\mu \, \Delta t$.) The variance of this distribution is of course $\sigma_V^2 = \mu^{-2} = \rho^2/\lambda^2$. Substituting in (11.11) thus gives

$$E(Q) = \frac{\rho}{1 - \rho}, \quad (11.12)$$

when the service-time is distributed in a negative exponential.

In the second case we have $\sigma_V{}^2 = 0$. So

$$E(Q) = \frac{\rho(2 - \rho)}{2(1 - \rho)} \tag{11.13}$$

for a constant service-time.

It is obvious that if the traffic intensity ρ is greater than unity, then customers are on average arriving faster than the rate at which they are being served: a queue is bound to build up, and we cannot expect to achieve equilibrium. If, however, $\rho = 1$ we might expect a balance to be maintained between the supply of and demand for the service in question. The general equation (11.11), as well as the special cases (11.12) and (11.13), shows that this is not so. With $\rho = 1$ we have $E(Q) = \infty$. In general, therefore, we must have a traffic intensity that is *less* than unity. This is a simple, though fundamental, property of queueing processes which is of considerable practical importance. We have only demonstrated this result on the assumption of random arrivals, though it is true in less restricted circumstances. It is obvious, of course, that in the very special case of a constant service-time, and regular arrivals at intervals equal to the service-time, we should have an equilibrium with $\rho = 1$. But as soon as some degree of statistical variation is introduced into the system we must have $\rho < 1$ for an equilibrium to be possible.

The foregoing type of argument can also be used in the case of random arrivals to obtain the average waiting-time for a customer. Suppose, for example, that a particular customer waits for a time W before he starts to receive attention from the server. If his service-time is V, he is in the system for a total time $W + V$. Let Q customers form a line behind him during this time. They will do so at an average rate of $\lambda(W + V)$, given W and V. Therefore

$$\lambda E(W + V) = E(Q)$$

$$= \frac{\rho(2 - \rho) + \lambda^2 \sigma_V{}^2}{2(1 - \rho)},$$

using (11.11). Since from (11.9), $E(V) = \rho/\lambda$, it follows that

$$E(W) = \frac{\rho^2 + \lambda^2 \sigma_V{}^2}{2\lambda(1 - \rho)}. \tag{11.14}$$

For maximum efficiency we should need $\sigma_V{}^2 = 0$, i.e. constant service-time. The mean waiting-time would then be

$$E(W) = \frac{\rho^2}{2\lambda(1 - \rho)}, \tag{11.15}$$

for a constant service-time.

In the negative-exponential case we should have, as before, $\sigma_V{}^2 = \rho^2/\lambda^2$. Substituting in (11.4) thus gives

$$E(W) = \frac{\rho^2}{\lambda(1 - \rho)}, \tag{11.16}$$

for a negative exponential service-time. Note that the average waiting-time given in (11.16) is precisely twice that appearing in (11.15).

Distributions of queue length and waiting-time

All the above results have been derived by nothing more complicated than an elementary use of the appropriate expectations. It is an interesting fact that these methods can easily be modified to yield expressions for the whole distributions of queue length and waiting-time. Notice that the latter, although possibly continuous for $W > 0$, will have a non-zero probability at $W = 0$.

Now we can write the probability-generating function for the equilibrium distribution of queue length (at any chosen epoch) as

$$P(x) = E(x^Q), \tag{11.17}$$

or, alternatively, as

$$P(x) = E(x^{Q+R-1+\delta}), \tag{11.18}$$

in virtue of (11.4).

Next, from (11.5) and (11.6), we have

$$P(\delta = 0) = \rho, \qquad P(\delta = 1) = 1 - \rho. \tag{11.19}$$

Using (11.19), we can split the expectations in (11.17) and (11.18) into two parts, namely,

$$P(x) = \rho \mathop{E}_{(\delta=0)} (x^Q) + (1 - \rho) \mathop{E}_{(\delta=1)} (x^Q)$$

$$= \rho x \mathop{E}_{(\delta=0)} (x^{Q-1}) + (1 - \rho), \tag{11.20}$$

and

$$P(x) = \rho \mathop{E}_{(\delta=0)} (x^{Q+R-1}) + (1 - \rho) \mathop{E}_{(\delta=1)} (x^R). \tag{11.21}$$

If arrivals are at random, then R is a Poisson variable with parameter λV, for given V, and so

$$E(x^R) = \mathop{E}_{V} (e^{\lambda V(x-1)}) \equiv H(x), \quad \text{say.} \tag{11.22}$$

The function $H(x)$ can clearly be expressed in terms of the Laplace transform of service-time. For if the latter has frequency function $b(v)$, with transform $b^*(s)$ given by

$$b^*(s) = \int_0^\infty e^{-sv} b(v) \, dv \equiv \underset{V}{E}(e^{-sv}), \qquad (11.23)$$

then
$$H(x) = b^*(\lambda(1 - x)). \qquad (11.24)$$

Using (11.22), we can write (11.21) in the form

$$P(x) = \rho H(x) \underset{(\delta = 0)}{E}(x^{Q-1}) + (1 - \rho)H(x). \qquad (11.25)$$

Eliminating $\underset{(\delta = 0)}{E}(x^{Q-1})$ between (11.20) and (11.25) leads to

$$P(x) = \frac{(1 - \rho)(1 - x)}{1 - x/H(x)} = \frac{(1 - \rho)(1 - x)}{1 - \dfrac{x}{b^*(\lambda(1 - x))}}, \qquad (11.26)$$

which expresses the probability-generating function of the queue left behind by a departing customer in terms of the Laplace transform of the service-time.

We now suppose that the waiting-time has the distribution function $C(w)$, with the Laplace transform

$$c^*(s) = \int_0^\infty e^{-sw} \, dC(w). \qquad (11.27)$$

From the relation between Q and $W + V$, namely that Q is a Poisson variable with mean $\lambda(W + V)$, given W and V, it follows that

$$\underset{Q}{E}(x^Q) = \underset{W,V}{E}(e^{\lambda(W + V)(x - 1)}),$$

i.e.
$$P(x) = b^*(\lambda(1 - x))c^*(\lambda(1 - x)), \qquad (11.28)$$

assuming that W and V are independent.

Equating the right-hand sides of (11.26) and (11.28) gives

$$b^*(u)c^*(u) = \frac{(1 - \rho)u/\lambda}{1 - \dfrac{1 - u/\lambda}{b^*(u)}},$$

where we have written $u = \lambda(1 - x)$. A little rearrangement then yields

$$c^*(u) = \frac{1 - \rho}{1 - (\lambda/u)\{1 - b^*(u)\}}. \qquad (11.29)$$

This formula enables us to write down the transform of the waiting-time distribution, given the service-time distribution (or at least the latter's transform).

In the special case of a negative exponential distribution of service-time, i.e.

$$b(v) = \mu e^{-\mu v} \tag{11.30}$$

(where, as before, $\mu = \lambda/\rho$), we have

$$b^*(s) = \frac{\mu}{s + \mu} \tag{11.31}$$

and, substituting (11.31) in (11.29),

$$c^*(s) = (1 - \rho)\left\{1 + \frac{\lambda}{s + (\mu - \lambda)}\right\}. \tag{11.32}$$

It is easy to see that $c^*(s)$ in (11.32) is just the transform of a distribution consisting of two components, namely, a non-zero probability of $1 - \rho$ at $w = 0$ and a continuous element with the negative exponential distribution of density given by

$$\rho(\mu - \lambda)e^{-(\mu - \lambda)w}, \qquad w > 0. \tag{11.33}$$

In this case we also have a simple expression for the distribution of queue length. Substituting (11.31) in (11.26) gives

$$P(x) = \frac{1 - \rho}{1 - \rho x}, \tag{11.34}$$

so that the distribution of queue length left behind by a departing customer has the geometric form

$$p_r = (1 - \rho)\rho^r, \qquad r = 0, 1, 2, \cdots. \tag{11.35}$$

11.3   Queues with many servers

Of the large number of factors that have been introduced into queueing models, one of the most important is the presence of several servers instead of only one, as supposed in the last section. Historically, the first problems to be studied were those arising in connection with a telephone exchange having a certain fixed number of channels. Fresh calls could be accepted until all channels were occupied. After that a waiting-line of calls would form. With a single channel we should have, in the terminology we have been using, only one server. With s channels, the queueing model would involve s servers. A typical application in the context of organizing medical care is to the problem of how many inpatient beds should be provided in any particular specialty. This is of particular importance where there is an acute shortage of hospital accommodations, e.g. post-war Britain, and waiting-lists for admissions are liable to be of an appreciable size.

It is clear that if accommodation is inadequate, patients will have to wait for admission. If it is lavish, on the other hand, the demand will outstrip the supply only very rarely, but the over-all economic cost may be unduly high. In order to see how to achieve a certain balance between these undesirable extremes, the problem can be formulated in queueing terms, so that the length of the waiting-list can be expressed as a function of the number of beds.

Suppose, to simplify matters, we assume that customers arrive at random at an average rate of λ per unit time, so that the chance of one new arrival in Δt is $\lambda \Delta t$. The distribution of the time interval between successive customers is therefore given by the negative exponential

$$f(u) = \lambda e^{-\lambda u}, \qquad 0 \leqslant u < \infty \tag{11.36}$$

(compare, for example, the arguments of Section 7.2).

We further assume that the service-time of any single customer also follows a negative exponential distribution, namely,

$$f(v) = \mu e^{-\mu v}, \qquad 0 \leqslant v < \infty. \tag{11.37}$$

This means that we may take the probability of departure during Δt, for any customer being served, as $\mu \Delta t$.

As usual, we may represent the chance of a total of n customers being present (including those being served) at time t by $p_n(t)$. If $n \leqslant s$ the chance of a customer departing in Δt is clearly $n\mu \Delta t$, since all customers can then be attended to simultaneously. But if $n \geqslant s$ the chance of a departure in Δt is fixed at $s\mu \Delta t$, because then exactly s customers are being served, while the remaining $n - s$ wait in line. We can derive the appropriate set of differential-difference equations for the probabilities by the standard method of relating $p_n(t + \Delta t)$ to $p_{n-1}(t)$, $p_n(t)$ and $p_{n+1}(t)$. We easily obtain (with $p_{-1} \equiv 0$)

$$\left.\begin{aligned}
\frac{dp_n(t)}{dt} &= \lambda p_{n-1}(t) - (\lambda + n\mu)p_n(t) + (n + 1)\mu p_{n+1}(t), \quad 0 \leqslant n < s \\
&= \lambda p_{n-1}(t) - (\lambda + s\mu)p_n(t) + s\mu p_{n+1}(t), \qquad s \leqslant n < \infty
\end{aligned}\right\}. \tag{11.38}$$

A full treatment of these equations for finite time is a matter of some complexity, but the equilibrium solution as $t \to \infty$, when it exists, is found relatively easily. For the equilibrium distribution p_n is given by $dp_n(t)/dt = 0$

for all n. We can then rearrange the right-hand side expressions in (11.38) in the form

$$0 = \lambda p_0 - \mu p_1 = \cdots = \lambda p_{n-1} - n\mu p_n = \cdots = \lambda p_{s-1} - s\mu p_s,$$

$$0 \leqslant n < s,$$

and

$$\lambda p_{s-1} - s\mu p_s = \lambda p_s - s\mu p_{s+1} = \cdots = \lambda p_{m-1} - s\mu p_m = \cdots, \quad s \leqslant m,$$

from which it quickly follows that

$$\left.\begin{aligned} p_n &= \left(\frac{\lambda}{\mu}\right)^n \frac{p_0}{n!}, & 1 \leqslant n < s \\ &= \left(\frac{\lambda}{\mu}\right)^n \frac{p_0}{s^{n-s}s!}, & s \leqslant n \end{aligned}\right\}. \tag{11.39}$$

Since $\sum_{n=0}^{\infty} p_n = 1$, we can determine p_0 from (11.39) as

$$p_0^{-1} = 1 + \sum_{n=1}^{s-1} \left(\frac{\lambda}{\mu}\right)^n \frac{1}{n!} + \frac{s^s}{s!} \sum_{n=s}^{\infty} \left(\frac{\lambda}{\mu s}\right)^n$$

$$= \sum_{n=0}^{s-1} \frac{(s\rho)^n}{n!} + \frac{(s\rho)^s}{(1-\rho)s!}, \tag{11.40}$$

where we now define the traffic intensity ρ as

$$\rho = \frac{\lambda}{s\mu}. \tag{11.41}$$

In order that (11.40) shall converge it is clearly necessary to have the condition $\rho < 1$.

From (11.39) and (11.40) we can calculate the average length of the queue and other indices of interest. For example, consider the number k of customers actually waiting to be served (excluding those being served). The average length of the line is given by

$$\begin{aligned} \bar{k} &= \sum_{n=s+1}^{\infty} (n-s)p_n \\ &= \sum_{n=s+1}^{\infty} \frac{(n-s)s^s\rho^n p_0}{s!} \\ &= \frac{(s\rho)^s p_0}{s!} \sum_{n=s+1}^{\infty} (n-s)\rho^{n-s} \\ &= \frac{\rho(s\rho)^s p_0}{(1-\rho)^2 s!}. \end{aligned} \tag{11.42}$$

The average time that a customer waits for service can also be found in a fairly simple form. If $n \leqslant s - 1$, a newly arrived customer finds a free server immediately. But if $n \geqslant s$, the customer must wait until $n - s + 1$ other customers have departed, taking on average $(n - s + 1)/s\mu$ units of time. The average waiting-time w is accordingly given by

$$\bar{w} = \sum_{n=s}^{\infty} \frac{n - s + 1}{s\mu} p_n$$

$$= \frac{(s\rho)^s p_0}{s\mu(s!)} \sum_{n=s}^{\infty} (n - s + 1)\rho^{n-s}$$

$$= \frac{(s\rho)^s p_0}{(1 - \rho)^2 s\mu(s!)}. \tag{11.43}$$

For computational purposes, expressions like (11.42) and (11.43) are most conveniently written in terms of Poisson probabilities and summations thereof. Thus the quantities

$$\left. \begin{array}{l} p(x, m) = e^{-m} m^x / x! \\[2mm] P(s, m) = \sum_{x=s}^{\infty} p(x, m) \end{array} \right\}, \tag{11.44}$$

and

have been tabulated in the tables by Molina (1942). We can, for example, easily rewrite (11.43) as

$$\bar{w} = 1 \bigg/ \left[s\mu(1 - \rho)\left\{ 1 + (1 - \rho)\frac{1 - P(s, s\rho)}{p(s, s\rho)} \right\} \right]. \tag{11.45}$$

Suppose we consider an application of the above results to a particular medical specialty. We will assume that the demand for admission as an inpatient is effectively at random, with an average of one patient per day, i.e. $\lambda = 1$. Next, we look at the service-mechanism. In many cases it is found that the length of stay is approximately negative exponential in character, though this is by no means always the case. Let us, therefore, make this assumption and take an average value of 20 days, i.e. $\mu = 1/20$. The traffic intensity depends of course on the number of servers available, that is on the number of beds in the ward. If the latter were actually 25 we should normally have fewer than this in effective use owing to the administrative difficulty of keeping beds fully occupied. In practice we might have an effective ward-size of $s = 22$ (this would imply a bed-occupancy of 88 per cent). With these figures we should have $\rho = 10/11$. Substitution in (11.45) gives $\bar{w} = 5 \cdot 7$ days. It is interesting to observe how small this figure is compared with the value $\bar{w} = \infty$ that occurs if $\rho = 1$. With slightly more accommodation, and a slightly smaller traffic intensity, the average waiting-time could be reduced to negligible proportions.

Of course, with fairly lavish accommodation the use of a queueing model is hardly necessary. But whenever a particular service is in short supply, it is of considerable benefit to be able to calculate in advance the likely practical consequences of any given degree of provision.

11.4   Monte Carlo methods in appointment systems

The kind of application discussed at the end of the last section, where we had a queue in the form of a waiting-list for entry to an inpatient bed, is only of practical importance when the available accommodation is of limited extent. Moreover, the salient features of the queueing sytem could be deduced from the long-term equilibrium theory of the appropriate model.

A somewhat different, though related, problem occurs in the conducting of outpatient sessions. Here, large queues may build up in the short run, even though the total amount of consultant's time is normally more than sufficient to deal with the demand, because a great many patients arrive at the beginning of the session. If there are of necessity a lot of unscheduled arrivals it may be difficult or impossible to find any means of cutting down the numbers of patients waiting for consultation. But if most patients come with an appointment to be seen at a particular clinic session, it is usually possible to achieve an optimal efficiency by carefully fixing the actual time of arrival as well.

Investigations in Britain (see Bailey, 1954, for references) showed that waiting-times of the order of one hour or more were all too frequent, and it was decided to employ queueing concepts to study the problem. It seemed that the best results would be obtained from a well-designed appointment system, such as those actually used in a few hospitals. If patients were given appointments at regular intervals, we should have a simple input-process for which the inter-arrival times were all equal to some fixed number. Apart from emergencies we could assume a "first come, first served" queue discipline. And the service mechanism would have to depend on a suitable distribution of consultation-time (regarded as covering the whole period during which the patient is in some way claiming the consultant's attention, or preventing him from attending to the next patient). In many cases it was found that a χ^2-distribution with about 5 degrees of freedom was typical.

Now, in principle, one could consider a number of alternative models with different inter-arrival times, and with variations in other features, such as the small block of patients which it is desirable to have at the beginning of a session, and then calculate the consequences of each in terms of waiting-times, etc. This would provide a means of discriminating between the different models, i.e. the different appointment systems.

Unfortunately, such a method entails dealing with a rather elaborate kind of queueing model in finite time. But even the mathematics (see next section) of the simplest model with random arrival and departure is fairly heavy. It was decided, therefore, to use an elementary form of Monte Carlo procedure. This is outlined briefly below.

Suppose, for the sake of illustration, we envisage a clinic in which 25 patients are to be seen, and the average consultation-time is 5 minutes. The session as a whole is then expected to last about $2\frac{1}{4}$ hours. In the studies referred to above a sampling experiment type of approach was adopted in which a standard statistical method was used to obtain 1250 random values from a χ^2-distribution with 5 degrees of freedom (the mean of 5 could conveniently be interpreted as being measured in minutes). These were then divided into 50 sets of 25, each set of 25 taken in some random order being regarded as the consultation times of 25 successive patients at a particular clinic. Although the total number of replicates, namely 50, was not very large, it was considered to yield results of the accuracy required for practical purposes.

A number of different appointment systems were then investigated, each system applied to all 50 hypothetical clinics. It is clearly a matter of simple arithmetic to calculate each patient's waiting time, and also the consultant's idle periods, i.e. the times he occasionally spends waiting for the next patient to arrive. In fact all calculations were done using ordinary desk computers, although it would be possible to obtain more accurate results by programing the work for an electronic computer and using far more than 50 replicates. The efficacy of each appointment system was judged according to the results achieved in terms of the patients' average waiting-time and the consultant's average idle period per clinic. Clearly, one of these can be made small only at the expense of making the other large. In practice therefore some kind of balance has to be struck.

There appeared to be some advantages in making the patients' inter-arrival time equal to the average consultation-time (there is nothing critical involved here in a finite, indeed short, queueing process). The main point at issue was how many patients ought to be present when the consultant started work. In some clinics it is customary to have quite a large number of patients in an initial block (this ensures long waits for patients but no idle periods for the consultant). The table below shows the values of the patients' average waiting-time and the consultant's average idle period per clinic for initial block sizes of 1 to 6.

It is clear from the table that quite a good practical balance is achieved by having an initial group of two patients only. The patients' average waiting-time is then 9 minutes, and the consultant's average idle period per clinic is only 6 minutes. This means in effect 6 minutes rest for the consultant in $2\frac{1}{4}$ hours, and a total of $9 \times 25 = 225$ minutes of waiting for patients. An initial block greater than two seems to give an undesirable lack of balance. If the consultant were 20 minutes late, this would of course have the effect of increasing the initial block from 2 to 6, with a very considerable increase in waiting-times.

Initial block of patients	Patients' average waiting-time (min.)	Consultant's average idle period per clinic (min.)
1	7	9
2	9	6
3	12	3
4	16	2
5	20	1
6	24	$\frac{1}{2}$

The table assumes an initial block of patients with subsequent arrivals at 5-minute intervals. There are 25 patients per clinic with an average consultation time of 5 minutes (χ_5^2-distribution).

There are of course many further important aspects of the above abbreviated discussion. And for a fuller account of these the reader should consult the literature cited. The main point to be made here is that the salient aspects of a real problem may often be investigated, at least in a preliminary way, by the use of even very simple Monte Carlo methods, when an exact mathematical analysis would prove to be intractable or excessively laborious.

11.5 Non-equilibrium treatment of a simple queue

So far in this chapter we have considered either the equilibrium distribution, when this exists, of various types of queueing situation, or the use of a Monte Carlo approach to a fairly complicated set-up in finite time. A full mathematical treatment in finite time is, for most queueing models, a matter of considerable complexity. In the present section we shall examine the comparatively simple case of a simple single-server queue with random

Poisson arrivals and a negative-exponential distribution of service-time (this is perhaps the simplest non-trivial case).

In essence, we have a population of individuals, whose numbers are augmented by Poisson arrivals, and diminished by Poisson departures, with the important proviso that departures can occur only if the population is greater than zero. The latter simple fact is the real source of the complications arising in the mathematical analysis. If we use the random-variable method of writing down a partial differential equation for, say, the probability-generating function, we must remember that so far as departures are concerned the expectation involved must omit the term arising from zero queue length.

To be specific, let the probability of a queue of length n at time t be $p_n(t)$. Suppose that the chance of an arrival in time Δt be $\lambda \, \Delta t$, and the chance of a departure be $\mu \, \Delta t$ if $n > 0$ and zero if $n = 0$. For reasons that will appear shortly it is convenient to work with a probability-generating function $P(z, t) = \sum_{n=0}^{\infty} p_n(t) z^n$, in which the dummy variable z is a complex quantity. In our usual notation there are two types of transition given by $j = 1$ and -1, with $f_1 = \lambda$ and $f_{-1} = \mu$. Remembering in the case of departures to omit a term corresponding to $n = 0$, we can immediately write down the required equation as

$$\frac{\partial P}{\partial t} = (z - 1)\lambda P + (z^{-1} - 1)\mu(P - p_0),$$

or
$$z \frac{\partial P}{\partial t} = (1 - z)\{(\mu - \lambda z)P - \mu p_0(t)\}, \tag{11.46}$$

with initial condition

$$P(z, 0) = z^a, \quad \text{say.} \tag{11.47}$$

Equation (11.46) is equivalent, on equating coefficients of z on both sides, to the system

$$\left.\begin{aligned}
\frac{dp_n}{dt} &= \lambda p_{n-1} - (\lambda + \mu)p_n + \mu p_{n+1}, \quad n \geq 1 \\[2mm]
\frac{dp_0}{dt} &= -\lambda p_0 + \mu p_1
\end{aligned}\right\}. \tag{11.48}$$

These equations can be obtained directly, by considering, as usual, $p_n(t + \Delta t)$ in terms of $p_n(t)$, etc., if any doubts are felt in the present context about writing down (11.46) immediately. The equilibrium solution of (11.48) has already been obtained by a different approach, and is given in (11.35) where $\rho = \lambda/\mu$.

It will be noticed that (11.46) contains the awkward term $p_0(t)$, which is an unknown function of t, on the right. This is an essential difference between this equation and those we have studied previously. The easiest way to handle the equation is to make use of the Laplace transform and its inverse, defined for any function $\phi(t)$ by the following relations

$$
\left.
\begin{aligned}
\phi^*(s) &= \int_0^\infty e^{-st}\phi(t)\, dt, \quad \mathbf{R}(s) > 0 \\[2mm]
\phi(t) &= \frac{1}{2\pi i}\int_{c-i\infty}^{c+i\infty} e^{st}\phi^*(s)\, ds
\end{aligned}
\right\}
\tag{11.49}
$$

where c is positive and greater than the real parts of all singularities of $\phi^*(s)$. (For an alternative treatment, see Cox and Smith, 1961, Section 3.1.)

Applying (11.49) to both sides of (11.46), gives after a little rearrangement

$$
P^*(z, s) = \frac{z^{a+1} - \mu(1 - z)p_0^*(s)}{sz - (1 - z)(\mu - \lambda z)},
\tag{11.50}
$$

where we have made use of the easily obtained result for a differential coefficient

$$
\left(\frac{d\phi}{dt}\right)^* = -\phi(0) + s\phi^*(s).
\tag{11.51}
$$

Apart from the fact that (11.50) contains the as yet unknown function $p_0^*(s)$, we have obtained an expression for $P^*(z, s)$ quite easily without having to solve a partial differential equation.

Now it follows from the definition of the Laplace transform of $P(z, t)$ that $P^*(z, s)$ must converge everywhere within the unit circle $|z| = 1$, provided $\mathbf{R}(s) > 0$. In this region, therefore, the zeros of both numerator and denominator on the right of (11.50) must coincide, and the corresponding factors cancel. Now the denominator vanishes at the points

$$
\left.
\begin{aligned}
\xi(s) &= \frac{(\lambda + \mu + s) - \{(\lambda + \mu + s)^2 - 4\lambda\mu\}^{\frac{1}{2}}}{2\lambda} \\[2mm]
\eta(s) &= \frac{(\lambda + \mu + s) + \{(\lambda + \mu + s)^2 - 4\lambda\mu\}^{\frac{1}{2}}}{2\lambda}
\end{aligned}
\right\}
\tag{11.52}
$$

where, in particular,

$$
\left.
\begin{aligned}
\xi + \eta &= (\lambda + \mu + s)/\lambda, \qquad \xi\eta = \mu/\lambda \\[2mm]
s &= -\lambda(1 - \xi)(1 - \eta)
\end{aligned}
\right\}
\tag{11.53}
$$

and on the right of (11.52) we take the value of the square root for which the real part is positive. We can easily show that $\xi(s)$ is the only zero within $|z| = 1$, if $\mathbf{R}(s) > 0$.

For on $|z| = 1$ we have

$$|(\lambda + \mu + s)z| = |\lambda + \mu + s| > \lambda + \mu \geqslant |\lambda z^2 + \mu|.$$

Hence by Rouché's Theorem

$$(\lambda + \mu + s)z - (\lambda z^2 + \mu) \equiv sz - (1 - z)(\mu - \lambda z)$$

has the same number of zeros as $(\lambda + \mu + s)z$ inside $|z| = 1$, namely just one. This zero must be $\xi(s)$ since $|\xi(s)| < |\eta(s)|$.

It follows that the numerator on the right of (11.50), as well as the denominator, vanishes when $z = \xi$. Therefore

$$p_0^*(s) = \frac{\xi^{a+1}}{\mu(1 - \xi)}. \tag{11.54}$$

The transform of $p_0(t)$ is thus determined. And the expression for $P^*(z, s)$ in (11.50) can be written as

$$P^*(z, s) = \frac{z^{a+1} - (1 - z)\xi^{a+1}(1 - \xi)^{-1}}{-\lambda(z - \xi)(z - \eta)}, \tag{11.55}$$

where we have replaced the denominator of (11.50) by the equivalent $-\lambda(z - \xi)(z - \eta)$.

We could have obtained this result by working with a real x instead of the complex z. But in more difficult situations than the present one, the complex variable method is more powerful, partly because of the advantages conferred by Rouché's theorem in examining zeros. It therefore seemed worth while to introduce the technique at this stage.

The expression in (11.55) can easily be expanded in powers of z. But we must first divide through by the factor $z - \xi$. Straightforward manipulation then yields

$$P^*(z, s) = \frac{1}{\lambda\eta} \sum_{j=0}^{a-1} \left\{ \frac{(\xi/\eta)^j - 1}{(\xi/\eta) - 1} + \frac{(\xi/\eta)^j}{1 - \xi} \right\} \xi^{a-j} z^j$$

$$+ \frac{1}{\lambda\eta} \left\{ \frac{(\xi/\eta)^a - 1}{(\xi/\eta) - 1} + \frac{(\xi/\eta)^a}{1 - \xi} \right\} \sum_{j=a}^{\infty} \eta^{a-j} z^j. \tag{11.56}$$

This can be written more concisely by using the transformation

$$\xi = \rho^{-\frac{1}{2}} w^{-1}, \qquad \eta = \rho^{-\frac{1}{2}} w. \tag{11.57}$$

We then have

$$P^*(z, s) = \lambda^{-1} \sum_{j=0}^{\infty} \left\{ \frac{1}{w^{a+j}(w - \rho^{-\frac{1}{2}})} - \frac{1}{w^{a+j+1} - w^{a+j-1}} \right.$$

$$\left. + \frac{1}{w^{|a-j|+1} - w^{|a-j|-1}} \right\} \rho^{-\frac{1}{2}(a-j-1)} z^j. \tag{11.58}$$

A neater final result is obtained if we work with the transform of $\partial P/\partial t$, i.e.

$$\{\partial P/\partial t\}^* = -z^a + sP^*(s), \quad \text{using (11.51)},$$

$$= -z^a - \lambda(1 - \xi)(1 - \eta)P^*, \tag{11.59}$$

using (11.53). If we now substitute in (11.59) the value of P^* given by (11.58), and pick out the coefficient of z^j, i.e. $\{\partial p_j/\partial t\}^*$, we find after a little elementary algebra that the latter is the sum of six terms each proportional to an expression of the type

$$\frac{1}{w^{\nu+1} - w^{\nu-1}}, \quad \nu \geqslant 0, \tag{11.60}$$

for which the inverse transform is

$$\frac{1}{2\pi i} \int_{c-i\infty}^{c+i\infty} \frac{e^{st}\, ds}{w^{\nu+1} - w^{\nu-1}} = \frac{-\frac{1}{2}(\nu+1)}{2\pi i} \int_{c-i\infty}^{c+i\infty} \frac{e^{st}\, ds}{(\eta - \xi)\eta^\nu}$$

$$= \frac{e^{-(\lambda+\mu)t}\frac{1}{2}\{2(\lambda\mu)^{\frac{1}{2}}\}^{\nu+1}}{2\pi i} \int_{c-i\infty}^{c+i\infty} \frac{e^{\zeta t}\, d\zeta}{(\zeta^2 - \alpha^2)^{\frac{1}{2}}\{\zeta + (\zeta^2 - \alpha^2)^{\frac{1}{2}}\}^\nu}$$

$$= (\lambda\mu)^{\frac{1}{2}}e^{-(\lambda+\mu)t}I_\nu\{2(\lambda\mu)^{\frac{1}{2}}t\}, \tag{11.61}$$

where in the second line we have written

$$\zeta = \lambda + \mu + s, \quad \alpha^2 = 4\lambda\mu, \tag{11.62}$$

and have then used a standard result from a table of integral transforms (e.g. Erdélyi, 1954), I_ν being a Bessel function of the first kind with imaginary argument.

The net result is that, for all j, we have

$$\frac{dp_j}{dt} = \rho^{\frac{1}{2}(j-a)}e^{-(\lambda+\mu)t}\{-(\lambda+\mu)I_{a-j} + (\lambda\mu)^{\frac{1}{2}}I_{a-j-1}$$

$$+ (\lambda\mu)^{\frac{1}{2}}I_{a-j+1} + \lambda I_{a+j+2} - 2(\lambda\mu)^{\frac{1}{2}}I_{a+j+1} + \mu I_{a+j}\}, \tag{11.63}$$

the suppressed Bessel function arguments being $2(\lambda\mu)^{\frac{1}{2}}t$. The actual values of the $p_j(t)$ can now be obtained by integration using the initial condition $p_j(0) = \delta_{aj}$. Actually, (11.63) can be thrown in a more convenient form, as we can combine the first three terms, and integrate them explicitly, if we use the standard result

$$I_{n-1}(x) + I_{n+1}(x) = 2I_n'(x). \tag{11.64}$$

We then obtain

$$p_j(t) = \rho^{\frac{1}{2}(j-2)} e^{-(\lambda+\mu)t} I_{a-j}$$

$$+ \rho^{\frac{1}{2}(j-a)} \int_0^t e^{-(\lambda+\mu)\tau} \{\lambda I_{a+j+2} - 2(\lambda\mu)^{\frac{1}{2}} I_{a+j+1} + \mu I_{a+j}\} \, d\tau, \quad (11.65)$$

the suppressed Bessel function arguments being $2(\lambda\mu)^{\frac{1}{2}}t$ or $2(\lambda\mu)^{\frac{1}{2}}\tau$, as before.

In the special critical case $\rho = 1$, i.e. $\mu = \lambda$, the foregoing procedure can be repeated on the integrand on the right of (11.65) to give the explicit result

$$p_j(t) = e^{-2\lambda t} \{I_{a-j}(2\lambda t) + I_{a+j+1}(2\lambda t)\}, \quad \rho = 1. \quad (11.66)$$

Calculation of moments

In principle, we can investigate the moments of the distribution of queue length by substituting e^θ for z in (11.55) to give the transform of the moment-generating function

$$M^*(\theta, s) = \frac{e^{(a+1)\theta} - (1 - e^\theta)\xi^{a+1}(1 - \xi)^{-1}}{-\lambda(e^\theta - \xi)(e^\theta - \eta)}. \quad (11.67)$$

Picking out the coefficients of θ and θ^2 yields the transforms of the first and second moments about the origin as

$$\{\mu_1'\}^* = \frac{a}{s} + \frac{\lambda - \mu}{s^2} + \frac{\mu\xi^a - \lambda\xi^{a+1}}{s^2}, \quad (11.68)$$

$$\{\mu_2'\}^* = \frac{a^2}{s} + \frac{2a(\lambda - \mu) + (\lambda + \mu)}{s^2} + \frac{2(\lambda - \mu)^2}{s^3}$$

$$+ \frac{\mu\xi^a - \lambda\xi^{a+1}}{s^2} \left\{ \frac{2(\lambda - \mu)}{s} - 1 \right\}. \quad (11.69)$$

In the special case of $\mu = \lambda$ these results simplify considerably. For the mean we can write the transform of the differential coefficient in time as

$$\left\{ \frac{d\mu_1'}{dt} \right\}^* = s\{\mu_1'\}^* - \mu_1'(0)$$

$$= \frac{\lambda(\xi^a - \xi^{a+1})}{s}$$

$$= \frac{\eta^{-a} + \eta^{-a-1}}{\eta - \xi}, \quad (11.70)$$

using (11.53) with $\mu = \lambda$. Now the inverse transform of $\eta^{-a}(\eta - \xi)^{-1}$ is

$$\frac{1}{2\pi i} \int_{c-i\infty}^{c+i\infty} \frac{e^{st}\, ds}{(\eta - \xi)\eta^a} = \lambda e^{-2\lambda t} \int_{c-i\infty}^{c+i\infty} \frac{(2\lambda)^a e^{ut}\, du}{(u^2 - 4\lambda^2)^{\frac{1}{2}}\{u + (u^2 - 4\lambda^2)^{\frac{1}{2}}\}^a}$$

$$= \lambda e^{-2\lambda t} I_a(2\lambda t), \tag{11.71}$$

where we have written $u = 2\lambda + s$, and used a standard transform. Substituting in (11.70) and integrating then gives

$$\mu_1'(t) = a + \lambda \int_0^t e^{-2\lambda\tau}\{I_a(2\lambda\tau) + I_{a+1}(2\lambda\tau)\}\, d\tau, \tag{11.72}$$

a form which is quite suitable for computation, using for example the British Association Tables (1950, 1952) of Bessel functions.

Asymptotic values

It is possible to invert (11.68) to give an expression for $\mu_1'(t)$, involving a complicated real integral which can be evaluated by direct computation. Of more immediate interest is to gain some general insight into the nature of the queueing process by considering asymptotic values for large t.

When $\lambda < \mu$, there is, as we have seen, an equilibrium solution for which, ultimately,

$$\lim_{t \to \infty} \mu_1'(t) = \frac{\rho}{1 - \rho}, \quad \lambda < \mu. \tag{11.73}$$

When $\lambda = \mu$, we can turn to (11.72), using the asymptotic formula for the Bessel function

$$I_\nu(x) \sim \frac{e^x}{(2\pi x)^{\frac{1}{2}}}, \quad x \text{ large}. \tag{11.74}$$

The integration is then elementary, and the leading term for large t gives

$$\mu_1'(t) \sim 2\left(\frac{\lambda t}{\pi}\right)^{\frac{1}{2}}, \quad \lambda = \mu. \tag{11.75}$$

This very interesting result means that in the critical case, when $\rho = 1$, the average queue length is asymptotically proportional to the square root of the time.

When $\lambda > \mu$, we have to consider rather carefully the expression

$$\frac{1}{2\pi i} \int_{c-i\infty}^{c+i\infty} \frac{e^{st}(\mu\xi^a - \lambda\xi^{a+1})\, ds}{s^2}. \tag{11.76}$$

The integrand has branch points at

$$s_1 = -(\lambda^{\frac{1}{2}} + \mu^{\frac{1}{2}})^2, \qquad s_2 = -(\lambda^{\frac{1}{2}} - \mu^{\frac{1}{2}})^2,$$

and a single pole at the origin. (When $\lambda < \mu$ the pole is double.) It can be shown that the value of the integral is at most $O(1)$. The largest term in the asymptotic expansion obtained from the inversion of (11.68) is thus $(\lambda - \mu)t$, i.e.

$$\mu_1'(t) \sim (\lambda - \mu)t, \quad \lambda > \mu, \tag{11.77}$$

as we might expect.

It is perhaps worth mentioning here that a useful way of obtaining an asymptotic expansion for expressions like (11.76) is to try to replace the contour $\int_{c-i\infty}^{c+i\infty}$ by the loop $\int_{-\infty}^{(0+)}$, which starts at infinity on the negative axis, goes round the origin counter-clockwise and returns to the starting point. This is legitimate here, since if $s = Re^{i\theta}$,

$$\xi(s) \to 0 \quad \text{and} \quad s^{-2}(\mu\xi^a - \lambda\xi^{a+1}) \to 0 \quad \text{as } R \to \infty.$$

If $\lambda \neq \mu$ we can replace $\int_{-\infty}^{(0+)}$ by $\int_{-\infty}^{(s_1+)}$ and a small circle enclosing the origin, which gives an easily evaluated integral. Now putting $u = s - s_1$, we obtain an integral with contour $\int_{-\infty}^{(0+)}$ and an integrand which, apart from a factor e^{ut}, can be expanded in positive powers of u. Using the standard result

$$\frac{1}{2\pi i} \int_{-\infty}^{(0+)} e^{ut} u^{n-1} \, du = \frac{\Gamma(n)\sin n\pi}{\pi t^n} \tag{11.78}$$

then gives an expansion in negative powers of t.

Similar methods to the above can be employed to study the variance. We easily find, for the equilibrium case,

$$\lim_{t \to \infty} \mu_2(t) = \frac{\rho}{(1 - \rho)^2}, \quad \lambda < \mu. \tag{11.79}$$

The other asymptotic values are

$$\mu_2(t) \sim 2\left(1 - \frac{2}{\pi}\right)\lambda t, \quad \lambda = \mu, \tag{11.80}$$

and

$$\mu_2(t) \sim (\lambda + \mu)t, \qquad \lambda > \mu. \tag{11.81}$$

For further details of the investigation into asymptotic behavior see Bailey (1957).

11.6 First-passage times

It is often of importance in studying a stochastic process to investigate the *first-passage time* to some given state. So far we have paid only scant attention to this idea, and it is convenient to examine it in more detail in the present context. (See, however, the discrete theory in Section 4.3.) Actually, the chance of extinction by time t, i.e. $p_0(t)$, for a birth-and-death process is simply the distribution function of the time to extinction. And the latter is the first-passage time to the state $n = 0$. More generally, we could inquire about the distribution of the time taken for the population first to rise to some pre-assigned value.

In queueing applications, first-passage times are of importance in what is called "rush-hour" theory. In a period of heavy demand, we want to know how long it may take for the queue to grow to some specified level. Alternatively, when the rush is over, we want to know how long it will take for a long queue to fall to some low value or even to disappear entirely. A useful method, which can be applied not only to queues, but much more generally to any kind of process is as follows.

Suppose, for example, we consider the simple queueing process of the last section, and imagine that we have a large queue with $n = a$ at $t = 0$. We then ask about the first-passage time to the state $n = b\,(b < a)$. We are, of course, most likely to be interested in such a question when $\lambda < \mu$. Let us consider a modification of the actual process in which we arbitrarily make $n = b$ an absorbing state. So far as the first-passage time to the latter state is concerned the original process and the modified form clearly have the same properties. But the advantage of the modified process is that $p_b(t)$ is the distribution function of the first-passage time. This process is best studied by first writing down, in the usual way, the basic differential-difference equations. The equations for the original process were given in (11.48), and the modified version is easily found to be

$$
\left.
\begin{aligned}
\frac{dp_n}{dt} &= \lambda p_{n-1} - (\lambda + \mu)p_n + \mu p_{n+1}, \quad n \geqslant b + 2 \\[2mm]
\frac{dp_{b+1}}{dt} &= -(\lambda + \mu)p_{b+1} + \mu p_{b+2} \\[2mm]
\frac{dp_b}{dt} &= \mu p_{b+1}
\end{aligned}
\right\}, \qquad (11.82)
$$

with initial condition

$$
p_a(0) = 1. \qquad (11.83)
$$

Multiplying the nth equation in (11.82) by z^{n+1} and adding gives the following partial differential equation for the probability-generating function $P(z, t) = \sum_{n=b}^{\infty} p_n z^n$:

$$z\frac{\partial P}{\partial t} = (1 - z)(\mu - \lambda z)\{P - p_b(t)z^b\}, \qquad (11.84)$$

with $$P(z, 0) = z^a. \qquad (11.85)$$

Equation (11.84) is somewhat similar to (11.46). This time, however, we have the explicit appearance of the unknown function $p_b(t)$. We adopt a method of solution entirely analogous to that used for (11.46). First, taking Laplace transforms of both sides of (11.84), and using (11.85), we obtain

$$P^*(z, s) = \frac{z^{a+1} - (1 - z)(\mu - \lambda z)z^b p_b^*(s)}{sz - (1 - z)(\mu - \lambda z)}. \qquad (11.86)$$

The previous analyticity argument about zeros of the denominator and numerator coinciding within the unit circle can also be applied here. We have immediately

$$p_b^*(s) = \frac{\xi^{a-b+1}}{(1 - \xi)(\mu - \lambda \xi)} = s^{-1}\xi^{a-b}. \qquad (11.87)$$

Now if τ is the first-passage time to the state $n = b$, the distribution function of τ is $p_b(\tau)$, and the frequency function $f(\tau)$, say, is $dp_b/d\tau$. Hence the transform of the frequency function is given by

$$f^*(s) = \left\{\frac{dp_b}{dt}\right\}^* = -p_b(0) + sp_b^* = \xi^{a-b}. \qquad (11.88)$$

The inverse transform is easily evaluated to give

$$f(\tau) = \frac{1}{2\pi i}\int_{c-i\infty}^{c+i\infty} \frac{e^{st}\mu^{a-b}\, ds}{(\lambda\eta)^{a-b}}$$

$$= \frac{e^{-(\lambda+\mu)\tau}(\mu/\lambda)^{\frac{1}{2}(a-b)}}{2\pi i}\int_{c-i\infty}^{c+i\infty} \frac{e^{v\tau}\{2(\lambda\mu)^{\frac{1}{2}}\}^{a-b}\, dv}{\{v + (v^2 - 4\lambda\mu)^{\frac{1}{2}}\}^{a-b}}$$

$$= (a - b)(\mu/\lambda)^{\frac{1}{2}(a-b)}\tau^{-1}e^{-(\lambda+\mu)\tau}I_{a-b}\{2(\lambda\mu)^{\frac{1}{2}}\tau\}, \qquad (11.89)$$

where we have written $v = \lambda + \mu + s$ in the second line, and have used a standard result. With $a = 1$ and $b = 0$ we obtain the distribution of a busy period.

The distribution of $f(\tau)$ in (11.89) is of a rather complicated form, but we can obtain moments directly from the Laplace transform since this is a moment-generating function. In fact, it is obvious that $f^*(s) = M(-s)$.

Hence we can pick out the coefficients of s in ζ^{a-b} to give the moments of τ about the origin. After a little algebra we find, for $\mu > \lambda$,

$$
\left.\begin{aligned}
m(\tau) &= \frac{a - b}{\mu - \lambda} \\[2mm]
\sigma^2(\tau) &= \frac{(a - b)(\lambda + \mu)}{(\mu - \lambda)^3}
\end{aligned}\right\}. \tag{11.90}
$$

Thus the average time taken to drop from $n = a$ to $n = b$ is exactly what would be expected from a purely deterministic calculation.

In a similar way we can investigate the time taken for a queue to build up from $n = a$ to $n = c$ $(c > a)$. This time we have the set of differential-difference equations

$$
\left.\begin{aligned}
\frac{dp_c}{dt} &= \lambda p_{c-1} \\[3mm]
\frac{dp_{c-1}}{dt} &= \lambda p_{c-2} - (\lambda + \mu)p_{c-1} \\[3mm]
\frac{dp_n}{dt} &= \lambda p_{n-1} - (\lambda + \mu)p_n + \mu p_{n+1}, \quad 1 \leqslant n \leqslant c - 2 \\[3mm]
\frac{dp_0}{dt} &= -\lambda p_0 + \mu p_1
\end{aligned}\right\}, \tag{11.91}
$$

where we have now modified the original process by introducing an absorbing barrier at $n = c$. The probability-generating function $P(z, t) = \sum_{n=0}^{c} p_n z^n$ can easily be shown to satisfy

$$
z \frac{\partial P}{\partial t} = (1 - z)\{(\mu - \lambda z)(P - p_c(t)z^c) - \mu p_0(t)\}, \tag{11.92}
$$

with initial condition

$$
p_a(0) = 1, \tag{11.93}
$$

as before.

Taking transforms of (11.92) and using (11.93) gives

$$
P^*(z, s) = \frac{z^{a+1} - (1 - z)\{\mu p_0^*(s) + (\mu - \lambda z)z^c p_c^*(s)\}}{sz - (1 - z)(\mu - \lambda z)}. \tag{11.94}
$$

Note that since the series $P(z, t)$ terminates with the term involving z^c, it is a polynomial in z. Thus $P^*(z, s)$ is also a polynomial in z. The numerator on the right of (11.94) must accordingly vanish at both zeros of the denominator. This enables us to determine the two unknown functions p_0^* and p_c^*.

The first-passage time τ for arrival at $n = c$ has a frequency function $f(\tau)$ whose transform is given by

$$f^*(s) = \left\{\frac{dp_c}{dt}\right\}^* = sp_c^*(s)$$

$$= \frac{\lambda(\xi^{a+1} - \eta^{a+1}) - \mu(\xi^a - \eta^a)}{\lambda(\xi^{c+1} - \eta^{c+1}) - \mu(\xi^c - \eta^c)}. \qquad (11.94)$$

Expanding (11.94) in powers of s gives moments as before. In general these are very complicated. But the mean has a relatively simple form, namely.

$$\left.\begin{aligned} m(\tau) &= \frac{c - a}{\lambda - \mu} - \frac{\mu}{(\lambda - \mu)^2}\left\{\left(\frac{\mu}{\lambda}\right)^a - \left(\frac{\mu}{\lambda}\right)^c\right\}, \quad \mu \neq \lambda \\ &= \frac{(c - a)(c + a + 1)}{2\lambda}, \qquad\qquad\qquad \mu = \lambda \end{aligned}\right\}. \qquad (11.95)$$

PROBLEMS FOR SOLUTION

1. Generalize the single-server queue with random arrivals and negative exponential service-time, treated in Section 11.5, to the case where the chance of an arrival in time Δt is $\lambda_n \Delta t$ and the chance of a departure is $\mu_n \Delta t$ for a queue length of size n (with $\mu_0 \equiv 0$).

Derive the basic differential-difference equations for the probabilities $p_n(t)$.

2. Find the equilibrium distribution, if it exists, for the queueing process of Problem 1.

Suggest a necessary and sufficient condition for the existence of such a stationary distribution (valid at any rate for "reasonable" values of the λ_n and μ_n).

3. Suppose that in a single-server queue of the foregoing type new customers are discouraged by the sight of a long queue, so that $\lambda_n = \lambda/(n + 1)$ while $\mu_n \equiv \mu$.

Prove that the queue length has a Poisson distribution with parameter $\rho = \lambda/\mu$.

4. Consider a single-server queue with arrival and departure rates λ and μ, respectively, where there is a waiting space that will hold only N customers. Customers who arrive and find the space full leave again immediately without service and do not return later.

Find the equilibrium distribution of the queue length. What proportion of customers who arrive have to depart without being served?

5. Generalize the process of Problem 4 to the case where there are N servers

(e.g. a telephone exchange with N trunk lines and no facilities for holding sub-scribers who cannot be supplied with a line immediately on request).

Find the equilibrium distribution of queue length.

6. Suppose that the simple single-server queue of Section 11.5 is generalized so that $\lambda = \lambda(t)$ and $\mu = \mu(t)$, i.e. the arrival and departure rates are time-dependent. Show that equations (11.46) and (11.48) still hold.

Eliminate the quantity $p_0(t)$ by further differentiation with respect to z.

Alternatively, change the time scale to τ where

$$\tau(t) = \int_0^t \mu(v)\, dv,$$

and write

$$\rho(\tau) = \frac{\lambda(t)}{\mu(t)}, \qquad R(\tau) = \frac{1}{\tau}\int_0^\tau \rho(w)\, dw.$$

Show that if we put $q_n(\tau) = e^{\tau(1+R)}p_n(\tau)$, then the q_n satisfy

$$\frac{dq_n}{d\tau} = \rho q_{n-1} + q_{n+1}, \quad n \geq 1,$$

$$\frac{dq_0}{d\tau} = q_0 + q_1.$$

Finally, using the generating function

$$Q(z, \tau) = \sum_{n=0}^{\infty} q_n(\tau)(z - \tau)^n/n!,$$

prove that process can be represented by the hyperbolic partial differential equation.

$$\frac{\partial^2 Q}{\partial \tau \partial z} = \rho Q,$$

with suitable boundary conditions.

(This equation can be solved, though with some difficulty, using Riemann's method. See A. B. Clarke (1956).)

CHAPTER 12

Epidemic Processes

12.1 Introduction

One of the most important applications of stochastic processes in the area of biology and medicine has been to the mathematical theory of epidemics. This is of interest, not only because of the biological and epidemiological implications, but also because a more complicated type of process is involved than those considered hitherto. So far, the continuous-time processes with which we have dealt have all involved models in which the transition probabilities were at most linear functions of the population size. The resulting partial differential equations for probability-generating functions or moment-generating functions then turned out to be of a linear type which was frequently soluble, or at least sufficiently tractable to yield a number of useful properties. With epidemic models, on the other hand, the transition probabilities are usually non-linear functions of the population size. And this leads, even with models that are descriptively very simple, to mathematical analyses of considerable complexity.

Alternatively, we may in certain circumstances prefer to employ a discrete-time model, and represent the epidemic process by some suitably defined Markov chain. Such models, which are typical of the so-called chain-binomial type, have an especial importance in the attempt to fit theoretical descriptions to actual data. However, so far as the analysis of epidemic processes in large groups is concerned, the discrete-time models are rather intractable, and it is usual to rely on the insights provided by continuous-time models for investigating the behavior of epidemics in reasonably large groups.

In the present chapter we shall look only at some of the salient features of epidemic processes, but for an extensive treatment of the mathematical theory of epidemics and a substantial bibliography the reader should refer to the book by Bailey (1957), and also to the appropriate parts of Bartlett

(1960). Before embarking on the mathematical analysis in subsequent sections, it will be convenient first to review briefly the basic ideas involved in the epidemiology of infectious diseases.

To begin with we suppose that we have a group of susceptible individuals all mixing homogeneously together. One, or more, of this group then contracts a certain infectious disease which may in due course be passed on to the other susceptibles. In general we assume that after the receipt of infectious material, there is a *latent period* during which the disease develops purely internally within the infected person. The latent period is followed by an *infectious period*, during which the infected person, or *infective* as he is then called, is able to discharge infectious matter in some way and possibly communicate the disease to other susceptibles. Sooner or later actual symptoms appear in the infective and he is removed from circulation amongst the susceptibles until he either dies or recovers. This removal brings the infectious period effectively to an end (at least so far as the possibility of spreading the disease is concerned). The time interval between the receipt of infection and the appearance of symptoms is of course the *incubation period*.

Most manageable models involve certain special cases of the above rather general situation. Thus in the simplest continuous-time treatments, covered in Sections 12.2, 12.3, and 12.4, we assume that the latent period is zero, so that an infected individual becomes infectious to others immediately after the receipt of the infection. It is also highly convenient to assume that the length of the infectious period has a negative exponential distribution. Another quite reasonable assumption is that the chance of any susceptible becoming infected in a short interval of time is jointly proportional to the number of infectives in circulation and the length of the interval. This means that the chance of one new infection in the whole group in a short interval of time will be proportional to the product of the number of infectives and the number of susceptibles, as well as the length of the interval. We thus have a transition probability which is a non-linear function of the group size, as mentioned above. And this is the chief source of the subsequent difficulties in the appropriate mathematical analysis.

In the simplest discrete-time models, on the other hand, such as the chain-binomial theory of Section 12.5, we tend to work with a constant latent period, and an infectious period which is very short, and in the limit contracted to a point.

Certain relaxations in the above severe simplications are possible, though as a rule even small modifications in the direction of greater generality produce a very considerable increase in the difficulty of mathematical analysis

12.2 Simple epidemics

We shall first look at the simplest possible kind of continuous-time epidemic model, in which we have the susceptibles in a group liable to catch the current infection, but in which there is no removal from circulation by death, recovery, or isolation. Such a model might well be approximately true for certain mild infections of the upper respiratory tract, where there is a comparatively long interval of time between the infection of any individual and his actual removal from circulation. The bulk of the epidemic would then take place before anyone was removed.

Deterministic case

It is convenient to begin by examining the appropriate deterministic version of such a situation. Partly, because we can sometimes use a deterministic component as a first approximation to a full stochastic model if numbers are sufficiently large; and partly, because it is instructive to see how the stochastic mean in small groups differs from the corresponding deterministic value.

Let us consider a homogeneously mixing group of total size $n + 1$, containing initially n susceptibles and just one infective. Suppose that at time t there are x susceptibles and y infectives, so that $x + y = n + 1$. According to the ideas discussed in the previous section, it is reasonable to suppose that, in a deterministic model, the actual number of new cases (regarded as a continuous variable) in an interval Δt is $\beta xy\, \Delta t$, where β is the contact-rate. We can thus write

$$\Delta x = -\beta xy\, \Delta t,$$

and the process is thus described by the differential equation

$$\frac{dx}{dt} = -\beta xy = -\beta x(n - x + 1).$$

Or, changing the time scale to $\tau = \beta t$, we have

$$\frac{dx}{d\tau} = -x(n - x + 1), \tag{12.1}$$

with initial condition

$$x = n, \qquad \tau = 0. \tag{12.2}$$

The solution of (12.1), subject to (12.2) is easily obtained as

$$x = \frac{n(n + 1)}{n + e^{(n+1)\tau}} \tag{12.3}$$

Now the form in which public records are often compiled is a statistical return showing the numbers of new cases appearing each day. It is, therefore, appropriate to work with the *epidemic curve*, which measures the rate $-dx/d\tau$ at which new cases arise, namely

$$-\frac{dx}{d\tau} = xy = \frac{n(n+1)^2 e^{(n+1)\tau}}{\{n + e^{(n+1)\tau}\}^2}. \tag{12.4}$$

The epidemic curve in (12.4) is a symmetrical unimodal curve passing through the point $(0, n)$, and with its maximum occurring when $\tau = (\log n)/(n + 1)$. We thus have the characteristic property of epidemics in which the rate of occurrence of new cases rises steeply to start with, reaches a peak, and then declines, finally reaching zero.

Stochastic case

Let us now construct a stochastic model of the deterministic set-up just discussed. We can suppose as before that initially there are n susceptibles and one infective. We now use a random variable $X(t)$ to represent the number of susceptibles still uninfected at time t, where the probability that $X(t)$ takes the value r is $p_r(t)$. At time t there are therefore $X(t)$ susceptibles and $n - X(t) + 1$ infectives. Only one type of transition is possible, namely the occurrence of a new infection, and a reduction of one unit in the number of susceptibles. Let us follow the suggestion in the previous section that we should make the chance of a new infection in a short interval of time proportional to the product of the number of susceptibles, the number of infectives, and the length of the interval. The chance of an infection in time Δt can thus be written as $\beta X(n - X + 1)\,\Delta t$, where β is the contact-rate. If we change the time scale to $\tau = \beta t$, this chance becomes $X(n - X + 1)\,\Delta\tau$.

In the terminology of Section 7.4 we have a single transition given by $j = -1$ with $f_{-1} = X(n - X + 1)$. The partial differential equation for the probability-generating function $P(x, \tau)$ is accordingly obtained from (7.29) as

$$\frac{\partial P}{\partial \tau} = (x^{-1} - 1)\left(x\frac{\partial}{\partial x}\right)\left\{(n+1) - \left(x\frac{\partial}{\partial x}\right)\right\}P$$

$$= (1 - x)\left(n\frac{\partial P}{\partial x} - x\frac{\partial^2 P}{\partial x^2}\right), \tag{12.5}$$

with initial condition

$$P(x, 0) = x^n. \tag{12.6}$$

The corresponding equation for the moment-generating function is

$$\frac{\partial M}{\partial \tau} = (e^{-\theta} - 1)\left\{(n + 1)\frac{\partial M}{\partial \theta} - \frac{\partial^2 M}{\partial \theta^2}\right\}, \tag{12.7}$$

with initial condition

$$M(\theta, 0) = e^{n\theta}. \tag{12.8}$$

The equations (12.5) and (12.7) have so far resisted attempts to find solutions in a simple closed form. (See, however, the recent paper by Bailey, 1963.) As we shall see, it is possible to find explicit expressions for the state-probabilities and low-order moments, but these are of a very complicated form if n is not very small.

First, we consider the basic differential-difference equation for the probabilities $p_r(\tau)$. This can be deduced from (12.5) by picking out the coefficient of x^r on both sides, or we can use the usual argument for the value of $p_r(\tau + \Delta t)$. We easily obtain

$$\left.\begin{aligned}\frac{dp_r}{d\tau} &= (r + 1)(n - r)p_{r+1} - r(n - r + 1)p_r, \quad 0 \leqslant r \leqslant n - 1 \\ \frac{dp_n}{d\tau} &= -np_n\end{aligned}\right\}, \tag{12.9}$$

with initial condition

$$p_n(0) = 1. \tag{12.10}$$

In principle, the system of equations in (12.9) could be solved successively, starting with the equation for p_n. Such a method is, however, impracticable for obtaining a general solution. The easiest procedure is to adopt the Laplace-transform approach, already used in Chapter 11 for handling queueing processes.

Let us use the Laplace transform as defined in (11.49), and write $q_r(s)$ for the transform of $p_r(\tau)$, so that

$$q_r(s) \equiv p_r^*(s) = \int_0^\infty e^{-s\tau}p_r(\tau)\, d\tau. \tag{12.11}$$

Applying the transformation to the system of equations in (12.9) and using (12.10) gives the recurrence relations

$$\left.\begin{aligned}q_r &= \frac{(r + 1)(n - r)}{s + r(n - r + 1)}q_{r+1}, \quad 0 \leqslant r \leqslant n - 1 \\ q_n &= \frac{1}{s + n}\end{aligned}\right\}. \tag{12.12}$$

We can now easily obtain the explicit expression for q_r given by

$$q_r = \frac{n!(n-r)!}{r!} \prod_{j=1}^{n-r+1} \{s + j(n-j+1)\}^{-1}, \quad 0 \leqslant r \leqslant n. \quad (12.13)$$

In principle, we only have to use the inverse transformation on the expression in (12.13) to derive an explicit formula for $p_r(\tau)$. The right-hand side of (12.13) can be expanded as a sum of partial fractions involving terms like $\{s + j(n-j+1)\}^{-1}$ and $\{s + j(n-j+1)\}^{-2}$. The latter arises when there are repeated factors, which occurs if $r < \frac{1}{2}(n+1)$. On inversion, the partial fractions just mentioned lead to $e^{-j(n-j+1)\tau}$ and $\tau e^{-j(n-j+1)\tau}$, respectively. Unfortunately, considerable labor is required to obtain the required partial fraction expansions explicitly. If, for example, we take n to be even, and $r > \frac{1}{2}n$, we can after much elementary algebra show that

$$p_r(\tau) = \sum_{r=1}^{n-r+1} c_{rk}\, e^{-k(n-k+1)\tau}, \quad (12.14)$$

where
$$c_{rk} = \frac{(-)^{k-1}(n-2k+1)n!(n-r)!(r-k-1)!}{r!(k-1)!(n-k)!(n-r-k+1)!}. \quad (12.15)$$

When $r \leqslant \frac{1}{2}n$ there are repeated factors in (12.13), and the formula for $p_r(\tau)$ is appreciably more complicated (see Bailey, 1957, Section 5.211 for details).

For values of n that are at all large, these formulas for the individual probabilities are too involved for easy manipulation, and we might expect to obtain a certain amount of information about means and variances, etc., by suitable use of equation (12.7) for the moment-generating function. The usual technique of equating coefficients of θ on both sides leads to the set of equations

$$\left.\begin{aligned}
\frac{d\mu_1'}{d\tau} &= -\{(n+1)\mu_1' - \mu_2'\} \\[2mm]
\frac{d\mu_2'}{d\tau} &= +\{(n+1)\mu_1' - \mu_2'\} - 2\{(n+1)\mu_2' - \mu_3'\} \\[2mm]
\text{etc.}&
\end{aligned}\right\}, \quad (12.16)$$

where μ_k' is the kth moment about the origin. Unfortunately, we cannot solve these successively, as the first equation contains μ_2' as well as μ_1'. This is the difficulty already alluded to earlier (see, for instance, the end of Section 8.6), and it arises primarily because the transition probabilities are no longer linear functions of the population size, as in most of the previous birth-and-death processes. Of course, if we already know μ_1', we could easily calculate μ_2' from the first equation in (12.16), and then higher moments in succession. For an approximate method based on neglecting cumulants higher than a given order, see Section 15.4.

Actually, an explicit expression for $\mu_1'(\tau)$ can be obtained by using the definition

$$\mu_1'(\tau) = \sum_{r=1}^{n} rp_r(\tau),\qquad(12.17)$$

and replacing the $p_r(\tau)$ by the values given by formulas of the type given in (12.14). Extremely heavy algebra is involved, but the final exact result (due to Haskey, 1954) is

$$\mu_1'(\tau) = \sum_{r=1}^{} \left[\frac{n!}{(n-r)!(r-1)!}\left\{(n-2r+1)^2\tau + 2 - \right.\right.$$

$$\left.\left. - (n-2r+1)\sum_{w=r}^{n-r} w^{-1}\right\}e^{-r(n-r+1)\tau}\right],\qquad(12.18)$$

where r runs from one up to $\tfrac{1}{2}n$ if n is even, and up to $\tfrac{1}{2}(n+1)$ if n is odd, with the proviso that in the latter case the term corresponding to $r = \tfrac{1}{2}(n+1)$ is one-half that given by the formula. It is clear that even (12.18) is too complicated for easy manipulation in general, and it is highly desirable to have some suitable asymptotic form valid for large n (see, for example, the recent paper by G. Trevor Williams, 1963).

For values of n that are not too large we can use (12.18) to compute the epidemic curve given by $-d\mu_1'/d\tau$ for comparison with the deterministic value appearing in (12.4). Bailey (1957) gives numerical results for $n = 10$ and $n = 20$, while Haskey (1954) gives figures for $n = 30$. The main feature of the stochastic epidemic curve is that although its maximum occurs at approximately the same time as for the deterministic curve, it is an asymmetrical curve which rises more slowly and falls away more slowly than in the deterministic case.

Duration time

One aspect of the simple stochastic epidemic that turns out to be fairly tractable is the *duration time*, i.e. the time that elapses before all susceptibles have become infected. Now when there are j infectives and $n - j + 1$ susceptibles, the chance of a new infection in $\Delta\tau$ is $j(n - j + 1)\,\Delta\tau$. The interval τ_j between the occurrence of the jth infection and the $(j + 1)$th infection therefore has a negative-exponential distribution given by

$$f(\tau_j) = j(n - j + 1)e^{-j(n-j+1)\tau_j}.\qquad(12.19)$$

It is clear that the τ_j are all independently distributed. Moreover, the duration time T is given by

$$T = \sum_{j=1}^{n} \tau_j. \tag{12.20}$$

The rth cumulant κ_r of the distribution of T is the sum of the rth cumulants of the distributions of the τ_j. But the rth cumulant of the distribution in (12.19) is simply $(r-1)! \, j^{-r}(n-j+1)^{-r}$. Hence κ_r is given by

$$\kappa_r = (r-1)! \sum_{j=1}^{n} j^{-r}(n-j+1)^{-r}. \tag{12.21}$$

For small n we could compute the cumulants directly from (12.21). But for larger values it is preferable to use asymptotic expressions. It can be shown that, neglecting terms of relative order n^{-1},

$$\left. \begin{aligned} \kappa_1 &= \frac{2(\log n + \gamma)}{n+1} \\ \kappa_r &= \frac{2(r-1)!}{n^r} \zeta(r) \end{aligned} \right\}, \tag{12.22}$$

where γ is Euler's constant, and $\zeta(r)$ is the Riemann ζ-function.

From (12.22) we see that the coefficient of variation, $\kappa_2^{\frac{1}{2}}/\kappa_1$, is asymptotically $\pi/(2\sqrt{3} \log n)$, while the usual measures of skewness and kurtosis take the limiting forms

$$\lim_{n \to \infty} \gamma_1 = \lim_{n \to \infty} \frac{\kappa_3}{\kappa_2^{\frac{3}{2}}} = \frac{2^{\frac{1}{2}}\zeta(3)}{\{\zeta(2)\}^{\frac{3}{2}}} = 0 \cdot 806,$$

and

$$\lim_{n \to \infty} \gamma_2 = \lim_{n \to \infty} \frac{\kappa_4}{\kappa_2^{2}} = \frac{3\zeta(4)}{\{\zeta(2)\}^{2}} = 1 \cdot 200.$$

Thus appreciable skewness and departure from normality persist even for infinite n. For groups of only moderate size the coefficient of variation is of substantial magnitude, being 27% when $n = 20$. The implication is that quite large variations in epidemiological behavior would be expected in practice. One should not, therefore, too easily ascribe apparently unusual results to such special factors as abnormal virulence or infectiousness, when the variation observed could be purely statistical in origin.

12.3 General epidemics

As we have seen in the previous section, there are considerable difficulties in a theoretical handling of even a simple epidemic in which we have only infection and no removal. We shall now look at the more complicated general situation in which both of these latter possibilities are realized. Many of the important properties of the general stochastic epidemic remain to be discovered. Thus we have information about the conditions required for the build-up of an epidemic, and we have a way of handling the distribution of the total number of cases of disease that may occur. But little is known about the exact form of the epidemic curve, or the distribution of the duration time.

Deterministic case

It is advantageous, as with the simple epidemic, to start with a brief investigation of the deterministic model. Let us suppose that we have a group of total size n of whom, at time t, there are x susceptibles, y infectives, and z individuals who are isolated, dead, or recovered and immune. Thus $x + y + z = n$. If the contact-rate is β, there will be $\beta xy\,\Delta t$ new infections in time Δt. But in addition to the occurrence of new cases of disease we must now include the removal of infectives from circulation. Let this take place at a rate γ, so that the number of removals in time Δt is $\gamma y\,\Delta t$. The deterministic process is clearly described by the differential equations

$$\left.\begin{aligned}
\frac{dx}{dt} &= -\beta xy \\[2mm]
\frac{dy}{dt} &= \beta xy - \gamma y \\[2mm]
\frac{dz}{dt} &= \gamma y
\end{aligned}\right\}, \qquad (12.23)$$

with initial conditions $(x, y, z) = (x_0, y_0, 0)$ when $t = 0$. If, moreover, there is only a trace of infection present initially, so that y_0 is small, then $x_0 \doteq n$.

An exact treatment of the equations in (12.23) is possible, at least in parametric terms. But for our present purpose it will suffice to derive certain required properties by means of a simpler, approximate, method.

To begin with, it follows from the second equation in (12.33) that no epidemic can build up unless $x_0 > \gamma/\beta$. For if the initial number of infectives is to increase, we must have $[dy/dt]_{t=0} > 0$. It is convenient to define a *relative removal-rate* given by $\rho = \gamma/\beta$. Thus the value $\rho = x_0 \doteq n$ constitutes a threshold density of susceptibles. For densities below this value, an initial trace of infection will be removed at a faster rate than it can build up. But for densities above the threshold an increasing number of cases will result.

An approximate solution of (12.23) can readily be obtained as follows.

We can eliminate y from the first and third equations to give $dx/dz = -x/\rho$, which on integration yields

$$x = x_0 e^{-z/\rho}. \tag{12.24}$$

The third equation can be rewritten in the form

$$\frac{dz}{dt} = \gamma(n - x - z),$$

since $y = n - x - z$. Using (12.24), this becomes

$$\frac{dz}{dt} = \gamma(n - z - x_0 e^{-z/\rho}). \tag{12.25}$$

We now expand the right-hand side of (12.25) as far as the term in z^2 to give, approximately,

$$\frac{dz}{dt} = \gamma\left\{n - x_0 + \left(\frac{x_0}{\rho} - 1\right)z - \frac{x_0}{2\rho^2}z^2\right\}. \tag{12.26}$$

Equation (12.26) is soluble by standard methods, and we eventually obtain the epidemic curve as

where
$$\left.\begin{array}{l} \dfrac{dz}{dt} = \dfrac{\gamma\alpha^2\rho^2}{2x_0}\operatorname{sech}^2(\tfrac{1}{2}\alpha\gamma t - \phi) \\[2ex] \alpha = \left\{\left(\dfrac{x_0}{\rho} - 1\right)^2 + \dfrac{2x_0 y_0}{\rho^2}\right\}^{\frac{1}{2}} \\[2ex] \phi = \tanh^{-1}\dfrac{x_0 - \rho}{\alpha\rho} \end{array}\right\}. \tag{12.27}$$

The graph is a symmetrical bell-shaped curve, and corresponds to the fact that in many actual epidemics the number of new cases reported each day rises to a maximum and dies away. It will be noted that we are here defining the epidemic curve in terms of the rate at which cases are removed from circulation. This is the natural thing to do, since in practice we can only record infectives as symptoms appear and they are removed: we cannot observe the occurrence of actual infections.

Ultimately, when the epidemic is over we shall have $dz/dt = 0$. It follows immediately from (12.26) that, if $x_0 \doteq n$, the final value of z is

$$z_\infty \doteq 2\rho\left(1 - \frac{\rho}{x_0}\right). \tag{12.28}$$

This is, of course, the total size of the epidemic. There will, as we have seen be no true epidemic if $x_0 < \rho$. Suppose therefore that $x_0 > \rho$ and that

$$x_0 = \rho + v. \tag{12.29}$$

Substitution in (12.28) shows that

$$z_\infty \doteq \frac{2\rho v}{\rho + v} \doteq 2v, \tag{12.30}$$

if v is small compared with ρ. The result in (12.30) shows that the total size of epidemic is $2v$, so that the initial density of susceptibles $\rho + v$ is finally reduced to $\rho - v$, i.e. to a point as far below the threshold as it was originally above it. This is Kermack and McKendrick's Threshold Theorem. A rather more precise version is possible by following through an exact analysis of (12.23), but the epidemiological implications are very similar. It will be interesting to compare this result with the stochastic analog developed below.

Stochastic case

We now consider the stochastic formulation of the general type of epidemic. There are, basically, two random variables to be taken into account here. As before, we can use $X(t)$ to represent the number of susceptibles still uninfected at time t. In addition, let $Y(t)$ be the number of infectives in circulation at time t. For a group of given total size, the number of individuals in the removed class is of course fixed once we know X and Y. The process is therefore two-dimensional, and may be treated, at least as far as developing the appropriate differential equations is concerned, as indicated in Chapter 10.

Let the probability of having u susceptibles and v infectives at time t be $p_{uv}(t)$, i.e.

$$P\{X(t) = u, \ Y(t) = v\} = p_{uv}(t). \tag{12.31}$$

It will also be convenient to work with the probability-generating function

$$P(z, w, t) = \sum_{u,v} p_{uv}(t) z^u w^v. \tag{12.32}$$

There are two kinds of possible transitions. First, there may be a new infection in Δt, which will occur with probability $\beta X Y \Delta t$, and u will decrease by one unit while v increases by one unit. Secondly, there may be a removal, which will occur with probability $\gamma Y \Delta t$, and v will decrease by one unit. In the terminology of Section 10.1 we have transitions represented by $(j, k) = (-1, +1)$ and $(0, -1)$; where $f_{-1, +1} = \beta X Y$, and $f_{0, -1} = \gamma Y$. We can then use (10.10) to write down immediately the required partial differential equation for $P(z, w, t)$, namely

$$\frac{\partial P}{\partial t} = \left\{ (z^{-1}w - 1)\beta \left(z \frac{\partial}{\partial z} \right) \left(w \frac{\partial}{\partial w} \right) + (w^{-1} - 1)\gamma \left(w \frac{\partial}{\partial w} \right) \right\} P$$

$$= \beta(w^2 - zw) \frac{\partial^2 P}{\partial z \partial w} + \gamma(1 - w) \frac{\partial P}{\partial w}. \tag{12.33}$$

If we change the time scale to $\tau = \beta t$, and write $\rho = \gamma/\beta$, as before, equation (12.33) becomes

$$\frac{\partial P}{\partial \tau} = (w^2 - zw) \frac{\partial^2 P}{\partial z \partial w} + \rho(1 - w) \frac{\partial P}{\partial w}. \tag{12.34}$$

Suppose the process starts at $\tau = 0$ with n susceptibles and a infectives. The initial condition is thus

$$P(z, w, 0) = z^n w^a. \tag{12.35}$$

So far no direct solution of (12.34) is available. But if we pick out the coefficients on $z^u w^v$ on both sides, or alternatively express $p_{uv}(t + \Delta t)$ in terms of the probabilities at time t, we obtain the set of differential-difference equations

$$\left. \begin{aligned} \frac{dp_{uv}}{d\tau} &= (u + 1)(v - 1)p_{u+1, v-1} - v(u + \rho)p_{uv} + \rho(v + 1)p_{u, v+1} \\ \frac{dp_{na}}{d\tau} &= -a(n + \rho)p_{na} \end{aligned} \right\}, \tag{12.36}$$

where $\quad 0 \leqslant u + v \leqslant n + a, \quad 0 \leqslant u \leqslant n, \quad 0 \leqslant v \leqslant n + a$

and $$p_{na}(0) = 1. \tag{12.37}$$

(We assume that any p_{uv}, whose suffices fall outside the permitted range, is zero.)

In principle we can handle these equations, as in the case of the simple epidemic equations in (12.9), by the Laplace-transform method. It is not difficult to see that we should finally be able to express each p_{uv} as a sum of terms involving exponential factors like $e^{-i(j+\rho)\tau}$. Unfortunately, no way has yet been found of expressing such results in a manageable closed form.

Suppose we use the Laplace transform in (11.49), writing $q_{uv}(s)$ for the transform of $p_{uv}(\tau)$, i.e.

$$q_{uv}(s) \equiv p_{uv}^*(s) = \int_0^\infty e^{-s\tau} p_{uv}(\tau) \, d\tau. \tag{12.38}$$

Equations (12.36) can then be transformed into

$$(u + 1)(v - 1)q_{u+1,v-1} - \{v(u + \rho) + s\}q_{uv} + \rho(v + 1)q_{u,v+1} = 0 \\ - \{a(n + \rho) + s\}q_{na} + 1 = 0 \right\}, \quad (12.39)$$

with the same ranges as before for u and v.

Total size of epidemic

Some progress can be made in the investigation of the total size of the epidemic. In fact, the total number of cases, not counting the initial ones, is the limit of $n - u$ as $t \to \infty$. It is easy to see that as $t \to \infty$, all the exponential terms in p_{uv} must vanish, leaving only a constant term, if any. Such a constant term will be the coefficient of s^{-1} in q_{uv}. Now the epidemic ceases as soon as $v = 0$. Hence the probability P_w of an epidemic of w cases in addition to the initial ones is given by

$$P_w = \lim_{t \to \infty} p_{n-w,0}, \quad 0 \leqslant w \leqslant n,$$

$$= \lim_{s \to 0} s q_{n-w,0}$$

$$= \lim_{s \to 0} \rho q_{n-w,1}, \quad (12.40)$$

where the last line follows from putting $v = 0$ in (12.39).

Thus we can write

$$P_w = \rho f_{n-w,1}, \quad 0 \leqslant w \leqslant n, \quad (12.41)$$

where

$$f_{uv} = \lim_{s \to 0} q_{uv}$$

with $\quad 1 \leqslant u + v \leqslant n + a, \quad 0 \leqslant u \leqslant n, \quad 1 \leqslant v \leqslant n + a \right\}. \quad (12.42)$

If we let $s \to 0$ in (12.39), and use (12.42), we obtain

$$(u + 1)(v - 1)f_{u+1,v-1} - v(u + \rho)f_{uv} + \rho(v + 1)f_{u,v+1} = 0 \\ - a(n + \rho)f_{na} + 1 = 0 \right\}. \quad (12.43)$$

These equations simplify to the recurrence formulas

$$g_{u+1,v-1} - g_{uv} + (u + \rho)^{-1}g_{u,v+1} = 0 \\ g_{na} = 0 \right\}, \quad (12.44)$$

by using the substitution

$$f_{uv} = \frac{n!(u + \rho - 1)!\,\rho^{n+a-u-v}}{vu!(n + \rho)!} g_{uv}. \quad (12.45)$$

If we solve (12.44) partially so as to express g_{uv} as a linear function of

$g_{u+1,k}$, $k = v - 1, \cdots, n + a - u - 1$, the following expressions result

$$
\left.
\begin{aligned}
g_{uv} &= \sum_{k=v-1}^{n+a-u-1} (u + \rho)^{v-k-1} g_{u+1,k}, \quad v \geqslant 2 \\
g_{u1} &= (u + \rho)^{-1} g_{u2} \\
g_{na} &= 1
\end{aligned}
\right\}
\tag{12.46}
$$

For small values of n we can use (12.46) to calculate algebraic expressions for the g_{uv}, and hence for the P_w. (Actually, we need only the g_{u1}, but the method entails computing all g_{uv}.) Bailey (1957) gives graphs of P_w for $n = 10, 20, 40$, with $a = 1$, for various values of ρ. It appears, that, as might be expected, for $\rho \geqslant n$ the distributions are all J-shaped with the highest point at $w = 0$. But if $\rho \ll n$ the distribution is U-shaped, so that there is *either* a very small number of total cases, *or* a large number of cases, intermediate situations being relatively rare. These results are obviously related to some kind of threshold phenomenon, but with groups as small as $n = 40$ there is no sharp transition from one type of distribution to the other.

An alternative method of handling the P_w is to construct a set of generating functions

$$
G_u(x) = \sum_{v=1}^{n+a-u} g_{uv} x^{v+1}, \quad 0 \leqslant u \leqslant n.
\tag{12.47}
$$

Multiplying the first equation in (12.44) by x^{v+2} and summing over v leads to

$$
G_u(x) = \frac{x^2}{x - (u + \rho)^{-1}} \{G_{u+1}(x) - (u + \rho)^{-1} g_{u1}\}, \quad 0 \leqslant u \leqslant n.
\tag{12.48}
$$

It is convenient to define

$$
G_{n+1}(x) \equiv x^a.
\tag{12.49}
$$

Now $G_u(x)$ must be a polynomial. Hence the denominator $x - (u + \rho)^{-1}$ on the right of (12.48) must also be a factor of the expression in braces, i.e. we must have

$$
g_{u1} = (u + \rho) G_{u+1}\left(\frac{1}{u + \rho}\right).
\tag{12.50}
$$

We can now use the recurrence relation in (12.48) repeatedly, with $x = (u + \rho)^{-1}$. After a certain amount of algebra we obtain the set of equations

$$
\sum_{w=0}^{j} \binom{n-w}{n-j} \left(\frac{n-j+\rho}{\rho}\right)^w P_w = \binom{n}{j} \left(\frac{n-j+\rho}{\rho}\right)^{-a}, \quad 0 \leqslant j \leqslant n.
\tag{12.51}
$$

These could be useful in further investigations of the total size of epidemic, but so far no explicit solutions are available

Yet another procedure is to adopt a random-walk approach. Thus the progress of the epidemic can be represented in terms of a succession of points (u, v). The random walk begins at (n, a) and ends at $(n - w, 0)$ for an epidemic of size w. The line $v = 0$ is clearly an absorbing barrier. There are, of course, two kinds of transitions with probabilities given by

$$\left. \begin{aligned} P\{(u, v) \to (u - 1, v + 1)\} &= \frac{u}{u + \rho} \\[2mm] P\{(u, v) \to (u, v - 1)\} &= \frac{\rho}{u + \rho} \end{aligned} \right\}. \tag{12.52}$$

We can now write down, more or less immediately, a formula for P_w by considering the sum of probabilities for all paths leading from (n, a) to $(n - w, 0)$. First consider all paths to $(n - w, 1)$ followed by a final step to $(n - w, 0)$. The required probability can be seen to be

$$P_w = \frac{n! \, \rho^{a+w}}{(n - w)! \, (n + \rho) \cdots (n + \rho - w)} \sum_\alpha (n + \rho)^{-\alpha_0} \cdots (n + \rho - w)^{-\alpha_w}; \tag{12.53}$$

where the summation is over all compositions of $a + w - 1$ into $w + 1$ parts such that

$$0 \leqslant \sum_{j=0}^{i} \alpha_j \leqslant a + i - 1,$$

for $0 \leqslant i \leqslant w - 1$ and $1 \leqslant \alpha_w \leqslant a + w - 1$. Unfortunately, if n is of appreciable size, there is some difficulty in using this result owing to uncertainty as to whether all relevant terms of the partition have been included.

Stochastic Threshold Theorem

In the purely deterministic case we saw that the Kermack and McKendrick Threshold Theorem provided a result of possible importance to epidemiological practice. We now inquire about the analogous result, if any, for the corresponding stochastic model. The properties of the total size of epidemic, referred to above, namely a J-shaped distribution for $\rho \geqslant n$ and a U-shaped distribution for $\rho \ll n$, when n is of moderate size, suggest that the value $\rho = n$ might provide a critical level at least for sufficiently large n.

That this is probably so can be seen without a detailed analysis. For if n is large, the population of infectives is initially subject, approximately, to a birth-and-death process in which the birth- and death-rates are βn and γ, respectively. (We have avoided the difficulty of non-linear transition probabilities simply by assuming that n is approximately constant, at least near the beginning of the epidemic.) Now, using the results in (8.59) with $\lambda = \beta n$ and $\mu = \gamma$, we see that chance of extinction for such a birth-and-death process is unity if $\rho \geqslant n$, and $(\rho/n)^a$ if $\rho < n$. Thus in the former case we expect only a small outbreak of disease, while in the latter we expect either a minor outbreak with probability $(\rho/n)^a$ or a major build-up with probability $1 - (\rho/n)^a$.

A more precise investigation by Whittle (1955), using an approximation method in which the actual epidemic process was regarded as lying between two other processes, confirms the general truth of these heuristic indications. Whittle's technique was to consider epidemics in which not more than a given proportion i of the n susceptibles were eventually attacked. The probability π_i of this is

$$\pi_i = \sum_{w=0}^{n_i} P_w. \qquad (12.54)$$

Whittle compared the actual epidemic process, for which the chance of one new infection in time Δt was $\beta uv \, \Delta t$, with two other processes for which the corresponding chances were $\beta nv \, \Delta t$ and $\beta n(1 - i)v \, dt$. These latter processes are "fast" and "slow" birth-and-death processes, for which explicit solutions are available. The final results can be expressed by saying that, for large n, there is zero probability of an epidemic exceeding any arbitrarily chosen intensity i if $\rho \geqslant n$; while if $\rho < n$, the chance of a true epidemic was $1 - (\rho/n)^a$ for small i.

12.4 Recurrent epidemics

It is a characteristic feature of many infectious diseases that each outbreak has the kind of epidemic behavior investigated in the last section, but in addition these outbreaks tend to recur with a certain regularity. The disease is then, in a sense, endemic as well as epidemic. It is of considerable interest to explore such a situation mathematically, and, as we shall see, there are some fundamental distinctions between the properties of deterministic and stochastic models.

Deterministic case

First, we consider a deterministic model which is practically the same as that at the beginning of Section 12.3, except for the modification that the stock of susceptibles is continually being added to at a constant rate μ, i.e. $\mu \, \Delta t$ new susceptibles are introduced into the population in time Δt. In order to keep the total population size constant the influx of new susceptibles must be balanced by an appropriate death-rate. Let us suppose that the latter affects only individuals in the removed group. We can then concentrate attention on the number of susceptibles and infectives, for whom the appropriate differential equations are easily seen to be

$$
\left.\begin{aligned}
\frac{dx}{dt} &= -\beta xy + \mu \\
\frac{dy}{dt} &= \beta xy - \gamma y
\end{aligned}\right\}. \tag{12.55}
$$

It is clear from (12.55) that an equilibrium state (x_0, y_0) is possible, and is given by putting $dx/dt = 0 = dy/dt$. We thus have

$$
x_0 = \gamma/\beta, \qquad y_0 = \mu/\gamma. \tag{12.56}
$$

The equations for small departures from the equilibrium value can be derived in the usual way by writing

$$
x = x_0(1 + u), \qquad y = y_0(1 + v), \tag{12.57}
$$

where u and v are small. Substituting (12.57) in (12.55) gives

$$
\left.\begin{aligned}
\sigma \frac{du}{dt} &= -(u + v + uv) \\
\tau \frac{dv}{dt} &= u(1 + v)
\end{aligned}\right\}. \tag{12.58}
$$

where $\qquad \sigma = \gamma/\beta\mu, \qquad \tau = 1/\gamma$.

We now neglect the product uv, and eliminate u from the two equations in (12.58), to give the second-order differential equation in v

$$
\frac{d^2v}{dt^2} + \frac{1}{\sigma}\frac{dv}{dt} + \frac{v}{\sigma\tau} = 0. \tag{12.59}
$$

This equation has a solution of type

$$
\left.\begin{aligned}
v &= v_0 e^{-t/2\sigma} \cos \xi t \\
\xi^2 &= \frac{1}{\sigma\tau} - \frac{1}{4\sigma^2}
\end{aligned}\right\}. \tag{12.60}
$$

where

We then obtain for u the result

$$
\left.\begin{aligned}
u &= v_0(\tau/\sigma)^{\frac{1}{2}}e^{-t/2\sigma} \cos(\xi t + \psi) \\
\cos \psi &= -\tfrac{1}{2}(\tau/\sigma)^{\frac{1}{2}}, \quad 0 \leqslant \psi \leqslant \pi
\end{aligned}\right\}. \tag{12.61}
$$

where

The important thing about these solutions is that they involve trains of *damped* harmonic waves. In his original investigation of measles outbreaks in London, Soper took τ equal to two weeks, the approximate incubation period. The period $2\pi/\xi$ for the available data was about 74 weeks, and σ about 68. This gives a peak-to-peak damping factor of $\exp(-\pi/\sigma\xi^{\frac{1}{2}}) \doteq 0.6$. Thus, although the constant introduction of new susceptibles is sufficient to account in some degree for the existence of epidemic waves, the latter are always damped out eventually to a steady endemic state. And this is precisely what is *not* observed. Successive outbreaks may vary somewhat in magnitude, but there is certainly no observable damping. We therefore turn to a stochastic formulation to see if this can serve us better.

Stochastic case

In constructing a stochastic analog of the foregoing deterministic model, we can start with the general stochastic epidemic of Section 12.3 and modify this by having new susceptibles introduced into the population according to a Poisson process with parameter μ. This means that there are now three possible types of transition in time Δt. As before, we use $X(t)$ and $Y(t)$ to represent the numbers of susceptibles and infectives at time t, with

$$P\{X(t) = u, Y(t) = v\} = p_{uv}(t).$$

A new infection will occur in Δt with probability $\beta X Y \Delta t$, and u decreases by one unit and v increases by one unit. A removal will occur with probability $\gamma Y \Delta t$, and v then decreases by one unit. In addition, we may have the introduction of a new susceptible, with probability $\mu \Delta t$, and u will increase by one unit. In the notation of Section 10.1, the transitions are represented by $(j, k) = (-1, +1), (0, -1)$, and $(+1, 0)$; and $f_{-1, 1} = \beta X Y$, $f_{0, -1} = \gamma Y$, and $f_{1, 0} = \mu$. Substitution in formula (10.10) then gives the partial differential equation for the probability-generating function $P(z, w, t)$ as

$$\frac{\partial P}{\partial t} = \beta(w^2 - wz)\frac{\partial^2 P}{\partial z \partial w} + \gamma(1 - w)\frac{\partial P}{\partial w} + \mu(z - 1)P, \qquad (12.62)$$

with initial condition

$$P(z, w, 0) = z^n w^a. \qquad (12.63)$$

So far, this basic equation for the process has proved intractable. We could use it to write down the differential-difference equations for the individual probabilities p_{uv}, but these too have not yielded to investigation.

In spite of the above difficulties in handling the full stochastic model of a general type of epidemic with an influx of new susceptibles, certain fundamental properties of the process can be elucidated, at least approximately. We saw how the threshold behavior of a general epidemic in a large group could be studied to some extent by observing that if n were large then at least initially the group of infectives was subject to a birth-and-death process with birth- and death-rates βn and γ, respectively; and starting at $t = 0$ with a individuals. Suppose we ignore the loss of susceptibles due to infection, but take the arrival of new susceptibles into account by assuming these to follow a deterministic process with arrival-rate μ. The approximate birth- and death-rates for the group of infectives will then be $\beta(n + \mu t)$ and γ. We then have a birth-and-death process which is non-homogeneous in time.

Such a process has already been studied in Section 9.3, where we may take the birth- and death-rates $\lambda(t)$ and $\mu(t)$ to be

$$\lambda(t) = \beta(n + \mu t), \quad \mu(t) = \gamma. \tag{12.64}$$

The quantity $\rho(t)$, defined by (9.26), is

$$\rho(t) = \int_0^t \{\mu(\tau) - \lambda(\tau)\}d\tau = (\gamma - \beta n)t - \tfrac{1}{2}\beta\mu t^2. \tag{12.65}$$

Although a general treatment of this process is a little involved, we can easily handle the important aspect of extinction. Thus the chance of extinction $p_0(t)$, given by (9.34), can be written as

$$p_0(t) = \left(\frac{J}{1 + J}\right)^a, \tag{12.66}$$

with

$$J = \int_0^t \gamma e^{(\gamma - \beta n)\tau - \tfrac{1}{2}\beta\mu\tau^2}\, d\tau$$

$$= \zeta e^{\tfrac{1}{2}\zeta^2(f-1)^2} \int_{\zeta(f-1)}^\infty e^{-\tfrac{1}{2}u^2}\, du, \tag{12.67}$$

where

$$\zeta = \frac{\gamma}{(\beta\mu)^{\frac{1}{2}}}, \quad f = \frac{n\beta}{\gamma} = \frac{n}{\rho}. \tag{12.68}$$

When $\mu = 0$, we have of course the stochastic threshold result already discussed for a general epidemic, with a discontinuity in the chance of extinction at $\rho = n$. But if $\mu \neq 0$, there is no sharp cut-off at $\rho = n$. The probability of extinction will vary continuously, having the value

$$\left\{1 + \left(\frac{2}{\pi\gamma^2}\right)^{\frac{1}{2}}\right\}^{-a}$$

at $\rho = n$. If ρ is much larger than n, we shall have $f \rightarrow 0$. If ζ is large, J will be large, and $P \rightarrow 1$. The numerical value of ζ suggested by the deterministic analysis is $\zeta = \gamma/(\beta\mu)^{\frac{1}{2}} = (\sigma/\tau)^{\frac{1}{2}} = 5.84$. This is quite sufficient to make $1 - P$ negligibly small.

The consequences of the above discussions are as follows. In the stochastic model we have the important phenomenon of extinction, which gives the process an entirely different character from the deterministic model. Suppose we consider a relatively small community, in which new susceptibles either appear by actual birth or by introduction from outside (as in schools and classes). After a major outbreak of disease both the numbers of susceptibles and infectives will be low. At this point ρ/n is likely to be large, and extinction of infectives therefore highly probable. The population of susceptibles will continue to increase, and will reach a level where a new outbreak will easily occur if there are chance contacts with other infectives outside the community, or if these are allowed to join the group. We shall thus have a series of recurrent outbreaks of disease, but with no damping and no steady equilibrium as in the deterministic model.

The consequences of introducing new infectives into the group, in order to trigger off a fresh epidemic as soon as the group of susceptibles has increased sufficiently, can be investigated to some extent by further approximate arguments. Thus, suppose that new infectives arrive at random with rate ε, so that the chance of a new arrival in Δt is $\varepsilon \Delta t$. Suppose also that at time $t = 0$ the number of susceptibles is negligible, but that it increases at a deterministic rate μ, which is much larger than ε. We neglect, as before, the ultimate reduction in the number of susceptibles when the disease begins to spread. Under these assumptions the effects of each newly introduced case of disease are independent of one another.

Consider now the interval $(u, u + \Delta u)$. The chance of no new infective appearing is $1 - \varepsilon \Delta u$. The chance that a new infective will arrive is $\varepsilon \Delta u$, when there will be a total of μu susceptibles. Any outbreak of disease resulting from this one case will be extinguished with probability $P(u)$ given by

$$\left. \begin{aligned} P(u) &= \frac{\rho}{\mu u} = \frac{\gamma}{\beta \mu u} = \frac{\sigma}{u}, \quad u > \sigma \\ &= 1, \qquad\qquad\qquad u \leqslant \sigma \end{aligned} \right\}, \qquad (12.69)$$

where σ is defined as $\gamma/\beta\mu$, as in (12.58). Thus the chance that no epidemic builds up from events occurring in Δu is $1 - \varepsilon \Delta u + \varepsilon \Delta u \, P(u)$. The chance of no epidemic occurring up to time t is accordingly

$$\prod_{0 < u < t} \{1 - \varepsilon \Delta u + \varepsilon \Delta u P(u)\} \sim \exp\left[-\varepsilon \int_0^t \{1 - P(u)\}\, du\right]$$

$$= \exp\left[-\varepsilon \int_\sigma^t \left(1 - \frac{\sigma}{u}\right) du\right], \quad \text{using (12.69),}$$

$$= \left(\frac{t}{\sigma}\right)^{\varepsilon\sigma} e^{-\varepsilon(t-\sigma)}, \quad t > \sigma. \tag{12.70}$$

The lower limit of the integral in the last line but one above is σ, since $P(u) = 1$ when $0 < u < \sigma$ by virtue of (12.69).

If we now write $F(t)$ for the distribution function of time elapsing before the occurrence of a major epidemic, the quantity in (12.70) is precisely $1 - F(t)$. The corresponding frequency function, given by differentiating with respect to t, can therefore be written as

$$\left. \begin{array}{c} f(T) = k(T-1)T^{k-1}e^{-k(T-1)}, \quad T > 1 \\[4pt] \text{where} \qquad T = t/\sigma, \qquad k = \varepsilon\sigma = \varepsilon\gamma/\beta\mu \end{array} \right\} \tag{12.71}$$

This distribution has a mode at $T_m = 1 + k^{-\frac{1}{2}}$, and we might reasonably suppose that the mean $\overline{T}$ was approximately the same. It can be shown that $\overline{T}$ is relatively independent of k unless k is less than about 2. This means that the average renewal time $\bar{t}$ for major epidemics is proportional to σ, but comparatively insensitive to changes in ε if the latter is not too small. It should be clear, at least in general terms, how the stochastic formulation of the model for recurrent epidemics leads to a permanent succession of undamped outbreaks of disease, although not exhibiting a strict sequence of oscillations.

12.5 Chain-binomial models

Finally, we shall look briefly at the discrete-time type of epidemic model. Such models have been used quite successfully in the statistical fitting of certain epidemic theories to data relating to small groups such as families. But so far as the treatment of epidemics in large groups is concerned, the Markov chains involved have as yet proved intractable. Up to the present, practically all large-scale epidemic phenomena have been investigated solely by means of the continuous-time models discussed in the previous sections of this chapter. It is perhaps worth taking a quick look at the way in which the discrete-time models can be constructed, as future developments may enable them to be used as a basis for the investigation of the corresponding stochastic processes.

The main idea is that we have a fixed latent period, which may be used as the unit of time, and an infectious period which is contracted to a single point. Let us suppose that u_t is the number of susceptibles in the group just before time t; that v_t is the number of infected individuals just before time t who will actually become infectious at that instant. We further define a *chance of adequate contact* $p (= 1 - q)$, which is the probability of a contact at any time between any two specified members of the group sufficient to produce a new infection if one of them is susceptible and one infectious.

It follows from the above that the chance that any given susceptible will escape infection at time t is q^{v_t}, i.e. will have adequate contact with none of the v_t infectives. Thus $1 - q^{v_t}$ is the chance of adequate contact with at least one infective, and this is what is required for infection actually to occur. The conditional probability of v_{t+1} new infections taking place (who will become infectious at time $t + 1$) is therefore given by the binomial distribution

$$P\{v_{t+1}|u_t, v_t\} = \binom{u_t}{v_{t+1}}(1 - q^{v_t})^{v_{t+1}}q^{v_t u_{t+1}}, \qquad (12.72)$$

where, of course,

$$u_t = u_{t+1} + v_{t+1}.$$

The process develops in a series of binomial distributions like (12.72). Hence the name of *chain-binomial* process.

The above formulation is essentially that originally used by Reed and Frost. An alternative version, due to Greenwood, assumes that the chance of infection does not depend on the actual number of infectives in circulation, provided there is one or more. In this case the chance of a given susceptible being infected at time t is simply p, instead of the $1 - q^{v_t}$ used above. The Greenwood version of (12.72) is easily seen to be

$$P\{v_{t+1}|u_t, v_t\} = \binom{u_t}{v_t}p^{v_{t+1}}q^{u_{t+1}}. \qquad (12.73)$$

Plainly, the disease spreads through any group of susceptibles until at some stage no fresh infections are produced to transmit the disease at the next stage. We can use a quasi-partitional notation to represent the chain of events occurring in a particular epidemic, by simply recording the number of infectious individuals at each stage beginning with v_0 primary cases to start the process off at $t = 0$. Thus a chain can be represented by $(v_0, v_1, \cdots, v_k)$, where v_k is the last non-zero value of the sequence of v's. The probability of this chain is seen from (12.73) to be

$$P\{v_0, v_1, \cdots, v_k\} = \prod_{t=0}^{k} \binom{u_t}{v_{t+1}} p^{v_{t+1}} q^{u_{t+1}} \tag{12.74}$$

$$= \frac{u_0!}{v_1! \cdots v_k! u_{k+1}!} p^{\sum\limits_{i=1}^{k} v_i} q^{\sum\limits_{j=1}^{k+1} u_j},$$

for the simpler Greenwood case, and

$$P\{v_0, v_1, \cdots, v_k\} = \frac{u_0!}{v_1! \cdots v_k! u_{k+1}!} q^{\sum\limits_{j=0}^{k} v_j u_{j+1}} \prod_{i=0}^{k-1} (1 - q^{v_i})^{v_{i+1}}, \tag{12.75}$$

in the Reed–Frost model.

The expressions in (12.74) and (12.75) allow us to write down the likelihood of any given type of chain in terms of one unknown parameter p. If we have data based on a fairly large number of independent family groups, we can use maximum-likelihood estimation to determine p, and can then use goodness-of-fit methods to test the actual feasibility of the model. There are a good many modifications and developments of these ideas, taking into account, for instance, variations in p, errors in counting susceptibles, loss of immunity, etc. For an extensive discussion of those topics the reader should consult Bailey (1957, Chapter 6).

PROBLEMS FOR SOLUTION

1. Consider the simple stochastic epidemic of Section 12.2.
Show that the moment-generating function $M_T(\theta)$ of the duration time T is given by

$$M_T(\theta) = -\theta q_0(-\theta), \quad \theta < 0.$$

Hence obtain the result

$$M_T(\theta) = \prod_{j=1}^{n} \left\{ 1 - \frac{\theta}{j(n-j+1)} \right\}^{-1},$$

and show that the rth cumulant of the distribution is

$$\kappa_r = (r-1)! \sum_{j=1}^{n} j^{-r}(n-j+1)^{-r},$$

as obtained in equation (12.21) by a different method.

2. Prove that the stochastic mean number of susceptibles for the simple stochastic epidemic is given, for sufficiently small values of τ, by the convergent series

$$\mu_1' = n - n\tau - \frac{n(n-2)}{2!} \tau^2 - \frac{n(n^2 - 8n + 8)}{3!} \tau^3 - \cdots.$$

3. Obtain explicit expressions for the probability distribution of the total size of a general stochastic epidemic where initially there is one infective and three susceptibles (i.e. $a = 1$, $n = 3$). Sketch the form of the distribution for different values of the relative removal-rate ρ, e.g. $\rho = 1, 2, 3, 4, 5$.

4. Calculate the probabilities for the different types of chain in the Reed–Frost and Greenwood versions of chain-binomial epidemics (a) in groups with one initial case and two further susceptibles, and (b) in groups with one initial case and three further susceptibles.

Calculate the stochastic mean appropriate to each stage, and plot for various values of p.

Examine the probability distribution of the duration time, i.e. the number of stages occurring before the epidemic ceases.

5. Suppose that a Greenwood type of chain-binomial epidemic occurs in a group of total size n, of which there are a primary cases (i.e. $v_0 = a$). If $_aP_{nj}$ is the probability of a *total* of j cases occurring in the whole epidemic, prove that the recurrence relation

$$_aP_{nj} = \sum_{k=1}^{j-a} \binom{n-a}{k} p^k q^{n-a-k} {}_kP_{n-a,j-a}$$

holds, where $_aP_{na} = q^{n-a}$.

CHAPTER 13

Competition and Predation

13.1 Introduction

When the transition probabilities for a stochastic process are no more than linear functions of the population size the mathematical analysis usually turns out to be reasonably tractable. And this may well hold even if the process involves more than one random variable. But as soon as the transition probabilities are non-linear functions of the population size, difficulties of a new kind arise. This happens, for instance, with epidemics. As we saw in Section 12.2, a fair amount of information can be obtained about the properties of a simple stochastic epidemic, but the general stochastic epidemic of Section 12.3 is as yet only partially investigated.

There are a variety of biological problems involving some kind of *interaction* between two or more species, such as competition between two species for a limited food supply, or a prey–predator relationship in which one species is part of the food supply of the second species. The classical work on such problems, entailing a deterministic analysis, is largely due to Lotka (1925) and Volterra (1931). The phenomena included in studies of this type are often referred to under the heading of "the struggle for existence".

While a good deal can be done along deterministic lines, our problem is to inquire about the properties of the corresponding stochastic models. The transition probabilities are typically non-linear functions of the population sizes, and all the difficulties already encountered in epidemic theory are present with additional complications. Comparatively little has been done so far by way of a full stochastic investigation. However, some progress has been made (see Bartlett, 1960, Chapters 4 and 5, for a more detailed account) by starting with a deterministic model and trying to see what modifications would be required in the results if certain probability aspects were introduced at those points where they would be most likely to have an appreciable effect. Such speculative and heuristic methods may be very useful at an early stage in a difficult stochastic treatment, as they may indicate what kind of solutions we should look for and what kind of analytic approximations might be appropriate.

13.2 Competition between two species

Let us consider first the situation in which there are two species competing for a common food supply. Suppose we start by thinking of only one of the species, and deciding what the appropriate deterministic model would be. It would be reasonable to imagine that when the species was relatively small in numbers the growth would be of exponential type, as in the remark following equation (8.17). But as the population increases, there is difficulty in maintaining growth because of the "environmental pressure" due to lack of food, overcrowding, etc. This suggests that the logistic type of equation,

$$\frac{dn}{dt} = \{\lambda - f(n)\}n, \tag{13.1}$$

might be appropriate, where n is the population size at time t, the birth-rate is λ, and the death-rate is $f(n)$, an increasing function of n with $f(0) = \mu$. With $f(n) \equiv \mu$, we have of course

$$dn/dt = (\lambda - \mu)n, \quad \text{or} \quad n = n_0{}^{(\lambda-}e^{\mu)t},$$

a simple exponential growth.

With two populations, having sizes n_1 and n_2 at time t, we could represent the deterministic model by a pair of equations of the form

$$\left.\begin{aligned}\frac{dn_1}{dt} &= (\lambda_1 - f_1)n_1 \\[2mm] \frac{dn_2}{dt} &= (\lambda_2 - f_2)n_2\end{aligned}\right\}, \tag{13.2}$$

where
$$f_i \equiv f_i(n_1, n_2) = \mu_{i1}n_1 + \mu_{i2}n_2; \quad i = 1, 2; \tag{13.3}$$

so that we now have death-rates in each population that are some suitable linear functions of the two population sizes.

Certain features of this deterministic model can be derived as follows. To begin with, if there is an equilibrium solution it will be given by $dn_1/dt = 0 = dn_2/dt$, i.e.

$$\left.\begin{aligned}\mu_{11}n_1 + \mu_{12}n_2 &= \lambda_1 \\[2mm] \mu_{21}n_1 + \mu_{22}n_2 &= \lambda_2\end{aligned}\right\}. \tag{13.4}$$

The required solution is the point

$$
\left.
\begin{aligned}
(n_1, n_2) &= \left(\frac{\lambda_1 \mu_{22} - \lambda_2 \mu_{12}}{\mu_{11}\mu_{22} - \mu_{12}\mu_{21}}, \frac{\lambda_2 \mu_{11} - \lambda_1 \mu_{21}}{\mu_{11}\mu_{22} - \mu_{12}\mu_{21}} \right) \\
&= (A, B), \quad \text{say,}
\end{aligned}
\right\}
\tag{13.5}
$$

provided $\mu_{11}\mu_{22} \neq \mu_{12}\mu_{21}$.

To study the behavior of small departures from the equilibrium point (A, B), we write

$$
n_1 = A(1 + \xi), \qquad n_2 = B(1 + \eta),
\tag{13.6}
$$

where ξ and η are small. Substitution of (13.6) in (13.2) gives

$$
\left.
\begin{aligned}
\left(\frac{d}{dt} + A\mu_{11} \right)\xi + B\mu_{12}\eta &= 0 \\
A\mu_{21}\xi + \left(\frac{d}{dt} + B\mu_{22} \right)\eta &= 0
\end{aligned}
\right\}.
\tag{13.7}
$$

The nature of the equilibrium depends on the roots of the quadratic

$$
x^2 + (A\mu_{11} + B\mu_{22})x + AB(\mu_{11}\mu_{22} - \mu_{12}\mu_{21}) = 0,
\tag{13.8}
$$

and it will be stable if $\mu_{11}\mu_{22} > \mu_{12}\mu_{21}$.

The latter result can be proved by elementary considerations by carefully examining in the (n_1, n_2) plane the values of dn_1/dt and dn_2/dt in relation to n_1 and n_2. In particular, we need to consider the signs of dn_1/dt and dn_2/dt above and below the lines given by (13.4). In the unstable case, $\mu_{11}\mu_{22} < \mu_{12}\mu_{21}$, it is easy to show that the point (n_1, n_2) ultimately moves to one axis or the other, depending on where it starts. The importance of this result is that in the deterministic case which species survives depends crucially on the initial condition. Thus in a stochastic modification we should expect random variation near the beginning of the process to be a vital factor in determining the end result. Moreover, it will be possible for one species *or* the other to survive under identical initial conditions.

In practice the life-cycles of competing animals, notably insects in the experimental work of Park (1954, 1957), involve considerable complications. The reader should consult Bartlett (1960, Chapter 5) for further discussion of these details.

Let us finally, in this section, write down the basic partial differential equation for a fully stochastic version of the foregoing deterministic model. We can suppose that the population sizes of the two species at time t are represented by the random variables $X_1(t)$ and $X_2(t)$. The chance of a birth in time Δt in the first population can be taken as $\lambda_1 X_1 \Delta t$, and in the second population as $\lambda_2 X_2 \Delta t$. The corresponding chances of deaths would be $(\mu_{11}X_1^2 + \mu_{12}X_1X_2)\,\Delta t$ and $(\mu_{12}X_1X_2 + \mu_{22}X_2^2)\,\Delta t$. In the notation of Section 10.1, the possible transitions are $(j,k) = (+1, 0)$; $(-1, 0)$; $(0, +1)$; $(0, -1)$; with $f_{1,0} = \lambda_1 X_1$, $f_{-1,0} = \mu_{11}X_1^2 + \mu_{12}X_1X_2$, $f_{0,1} = \lambda_2 X_2$ and $f_{0,-1} = \mu_{21}X_1X_2 + \mu_{22}X_2^2$. The desired partial differential equation can now be written immediately, using (10.9), as

$$\frac{\partial M}{\partial t} = \lambda_1(e^{\theta_1} - 1)\frac{\partial M}{\partial \theta_1} + (e^{-\theta_1} - 1)\left(\mu_{11}\frac{\partial^2 M}{\partial \theta_1^2} + \mu_{12}\frac{\partial^2 M}{\partial \theta_1 \partial \theta_2}\right)$$

$$+ \lambda_2(e^{\theta_2} - 1)\frac{\partial M}{\partial \theta_2} + (e^{-\theta_2} - 1)\left(\mu_{21}\frac{\partial^2 M}{\partial \theta_1 \partial \theta_2} + \mu_{22}\frac{\partial^2 M}{\partial \theta_2^2}\right), \quad (13.9)$$

where, of course, $M \equiv M(\theta_1, \theta_2, t)$. Equation (13.9) is a more complicated type of second-order partial differential equation than those we have developed hitherto. Not surprisingly, therefore, no explicit solution of any kind is yet available. If we equate coefficients of powers of θ_1 and θ_2 on both sides of (13.9), the first two equations are

$$\left.\begin{aligned}
\frac{dm_{10}}{dt} &= \lambda_1 m_{10} - \mu_{11}m_{20} - \mu_{12}m_{11}\\[2mm]
\frac{dm_{01}}{dt} &= \lambda_2 m_{01} - \mu_{21}m_{11} - \mu_{22}m_{22}
\end{aligned}\right\}, \qquad (13.10)$$

where we have written m_{ij} for the (ij)th joint moment of X_1 and X_2 about the origin, i.e. $E(X_1) = m_{10}$, $E(X_2) = m_{01}$, etc.

Equation (13.10) is the stochastic analog of (13.2). Unfortunately, we cannot even begin to look for a solution since these first two equations already involve 5 unknowns. However, it is possible that the approximate procedure dealt with in Section 15.4 might be of service. The first 5 equations (2 first-order; 3 second-order) obtained in this way from the partial differential equation for the cumulant-generating function would contain only 5 unknowns, namely κ_{10}, κ_{01}, κ_{20}, κ_{11}, and κ_{02}, if cumulants of third and higher orders could be neglected.

13.3 A prey–predator model

A second situation, mentioned in the introductory section to this chapter, is where one species is at least part of the food supply of the other. We then have a prey–predator relationship. The first species can be regarded as the prey or host, and the second as the predator or parasite, according to circumstances.

The simplest deterministic model of the Lotka–Volterra type is based on the frequency of encounter between members of the two species. And it is reasonable, as a first approximation, to take this frequency as proportional to the population sizes. Thus in time Δt we can suppose that the first species gains an amount $\lambda_1 n_1 \Delta t$ due to new births and loses an amount $\mu_1 n_1 n_2 \Delta t$ due to predation. Similarly, if the second species has a birth-rate proportional to the number of prey available we can take its gain in Δt as $\lambda_2 n_1 n_2 \Delta t$, and its loss due to death as $\mu_2 n_2 \Delta t$. The appropriate differential equations are thus

$$\left. \begin{aligned} \frac{dn_1}{dt} &= \lambda_1 n_1 - \mu_1 n_1 n_2 \\ \frac{dn_2}{dt} &= \lambda_2 n_1 n_2 - \mu_2 n_2 \end{aligned} \right\}. \tag{13.11}$$

This time there is an equilibrium point given by

$$(n_1, n_2) = (\mu_2 / \lambda_2, \lambda_1 / \mu_1). \tag{13.12}$$

We can show that this equilibrium point is neutral in the sense that there is no damping towards it for population points at any other part of the (n_1, n_2) plane. Thus division of the two equations in (13.11) gives

$$\frac{dn_1}{dn_2} = \frac{(\lambda_1 - \mu_1 n_2)n_1}{(\lambda_2 n_1 - \mu_2)n_2}, \tag{13.13}$$

which integrates to give

$$f(n_1, n_2) \equiv -\mu_2 \log n_1 + \lambda_2 n_1 - \lambda_1 \log n_2 + \mu_1 n_2 = \text{constant}. \tag{13.14}$$

The solution is therefore represented by the set of closed curves given by $f(n_1, n_2) = \text{constant}$, for different values of the constant.

For small curves around the equilibrium point we can write

$$n_1 = \frac{\mu_2}{\lambda_2}(1 + \xi), \qquad n_2 = \frac{\lambda_1}{\mu_1}(1 + \eta), \tag{13.15}$$

where ξ and η are small. Retaining only second-order terms in ξ and η, we see that the paths are approximately the ellipses

$$\mu_2 \xi^2 + \lambda_1 \eta^2 = \text{constant}. \tag{13.16}$$

If larger curves are considered, it is clear that they will tend to be deformed from the elliptical shape by the presence of the two axes.

Let us now imagine how these results would be modified by the introduction of suitable probability elements. If the equilibrium point is sufficiently far from each axis, i.e. if μ_2 / λ_2 and λ_1 / μ_1 are large enough, the deterministic model would be reasonably satisfactory for paths on which n_1 and n_2 remained fairly large. But if either of these quantities became small enough for statistical fluctuations to be appreciable, a chance variation might cause the population point (n_1, n_2) to strike one of the axes, which are of course absorbing barriers. Hence a stochastic model will be inherently unstable, because sooner or later the stochastic drift will cause one of the species to die out. The latter may be the predator itself; or the prey, in which case the predator will die from lack of food.

Although the final extinction of the predator is thus certain on such a stochastic model, the chance of extinction after only one or two cycles of the population point about the equilibrium value may be quite small. An exact analysis of this situation is not available, but we can undertake an approximate discussion as follows. Consider the deterministic curves $f(X_1, X_2) = $ constant. Elementary examination of the equations in (13.16), especially with regard to the signs of the differential coefficients, shows that these curves are traversed in an anticlockwise sense. Let us now look at the behavior in the neighborhood of the equilibrium point. Substituting (13.15) in (13.11) gives

$$\frac{d\xi}{dt} = -\lambda_1\eta, \qquad \frac{d\eta}{dt} = \mu_2\xi,$$

for small ξ and η. These equations imply

$$\frac{d^2\xi}{dt^2} = -\lambda_1\mu_2\xi. \tag{13.17}$$

It immediately follows that the motion is periodic with period

$$T = 2\pi(\lambda_1\mu_2)^{-\frac{1}{2}}.$$

Let us now suppose that the prey is fairly abundant so that the chance of extinction in a stochastic modification of the deterministic model is small. This means that we regard the equilibrium value of n_1, namely μ_2/λ_2, as large enough for probability effects near $(\mu_2/\lambda_2, \lambda_1/\mu_1)$ to be negligible so far as the first species is concerned. If, however, n_2 is only of moderate size, statistical fluctuations may well be important. They will have their greatest effect when n_2 has its smallest value, n_2' say, at a point on the relevant elliptical path nearest to the n_1-axis. At this point $dn_2/dn_1 = 0$. Hence, from (13.13), $n_1 = \mu_2/\lambda_2$. Substituting the latter in the second equation of (13.11) shows that $dn_2/dt = 0$. The effective birth- and death-rates per individual in the second population, namely $\lambda_2 n_1$ and μ_2, are thus both equal at the point in question. But n_1 is increasing, and so we may write the birth-rate of n_2 as $\mu_2 + \gamma \sin(2\pi t/T)$; where T is the period $2\pi(\lambda_1\mu_2)^{-\frac{1}{2}}$; t is measured from the time of occurrence of the smallest value of n_2; and the amplitude γ depends on the actual path being followed.

Thus in the neighborhood of n_2' we can regard the second population as being approximately subject to a birth-and-death process with birth- and death-rates $\mu_2 + \gamma \sin(2\pi t/T)$ and μ_2, respectively. This is a non-homogeneous process. In the notation of Section 9.3 we have, from (9.26),

$$\rho(t) = \int_0^t \{\mu(\tau) - \lambda(\tau)\}\, d\tau$$

$$= -\int_0^t \gamma \sin(2\pi\tau/T)\, d\tau$$

$$= \frac{\gamma T}{2\pi}\left(\cos\frac{2\pi t}{T} - 1\right). \tag{13.18}$$

The chance of extinction, given by (9.34), can thus be written as

$$p_0(t) = \left(\frac{J}{1+J}\right)^{n_2'}, \tag{13.19}$$

where n_2' is the smallest value of n_2, and

$$J = \int_0^t \mu_2 e^{\rho(\tau)}\, d\tau. \tag{13.20}$$

The integral in (13.20) can be evaluated for any t, but we are specially interested in the region where t is of order $\tfrac{1}{4}T$. A comparatively simple solution exists for $t = T$, and we should not underestimate the value at $t = \tfrac{1}{4}T$ if we used $t = T$ instead, because the chance of extinction always increases with t. In any case we are mainly interested in orders of magnitude in discussions of the present type. When $t = T$, the integral in (13.20) is

$$J(T) = \int_0^T \mu_2 e^{\rho(\tau)}\, d\tau$$

$$= \mu_2 T e^{-\gamma T/2\pi} I_0(\gamma T/2\pi), \tag{13.21}$$

where I_0 is a Bessel function of the first kind, of zero order, and with imaginary argument.

Bartlett (1957, 1960) has given a detailed discussion of some data of Gause (1938) on the growth of *Paramecium aurelia* when feeding on yeast cells. It is doubtful whether any real ecological situation is sufficiently homogeneous for the simple kind of mathematical model we are discussing to be at all adequate. But, for what it was worth, the data suggested the values $\lambda_1 \sim 1$, $\mu_2 \sim 0.45$ (with the time measured in days), and an equi-librium point at about $(1.5 \times 10^7, 100)$. Thus $\mu_1 \sim 0.01$, $\lambda_2 \sim 3 \times 10^{-8}$. This gives the period of the cycle as 9.4 days. In Gause's data $\gamma \sim \frac{5}{8}\mu_2$. Hence $J \sim 2.6$. A typical value of n_2' was 15. Substitution of these quantities in (13.19) gives the chance of extinction at the critical phase of the cycle as of order $e^{-5.7} \sim 0.003$. Bartlett's discussion is also supported by simulation studies based on the Monte Carlo method. Although all those arguments are very tentative and heuristic in nature, they do provide some qualitative basis for an approximate evaluation of the stochastic aspects.

As in the previous section, there is little difficulty in writing down the partial differential equation for the moment-generating function appro-priate to a fully stochastic model of the prey–predator situation. Let the sizes of the two species are represented at time t by the random variables $X_1(t)$ and $X_2(t)$. The chance of a birth in the first population in time Δt is $\lambda_1 X_1 \, \Delta t$ and the chance of a death is $\mu_1 X_1 X_2 \, \Delta t$. The corresponding quantities for the second population are $\lambda_2 X_1 X_2 \, \Delta t$ and $\mu_2 X_2 \, \Delta t$. The usual functions f_{jk} are thus $f_{1,0} = \lambda_1 X_1$, $f_{-1,0} = \mu_1 X_1 X_2$, $f_{0,1} = \lambda_2 X_1 X_2$ and $f_{0,-1} = \mu_2 X_2$. Equation (10.9) now gives

$$\frac{\partial M}{\partial t} = \lambda_1(e^{\theta_1} - 1)\frac{\partial M}{\partial \theta_1} + \mu_1(e^{-\theta_1} - 1)\frac{\partial^2 M}{\partial \theta_1 \partial \theta_2}$$

$$+ \lambda_2(e^{\theta_2} - 1)\frac{\partial^2 M}{\partial \theta_1 \partial \theta_2} + \mu_2(e^{-\theta_2} - 1)\frac{\partial M}{\partial \theta_2}. \quad (13.22)$$

Equating coefficients of θ_1 and θ_2 on both sides gives

$$\left.\begin{aligned} \frac{dm_{10}}{dt} &= \lambda_1 m_{10} - \mu_1 m_{11} \\[2mm] \frac{dm_{01}}{dt} &= \lambda_2 m_{11} - \mu_2 m_{01} \end{aligned}\right\}, \quad (13.23)$$

which, although slightly simpler than the previous equations (13.10) for the competition model, still encounters the difficulty of involving two equations with three unknowns. The kind of approximation suggested at the end of the previous section might prove serviceable in handling this process, which otherwise seems to be mathematically intractable.

CHAPTER 14

Diffusion Processes

14.1   Introduction

In Chapters 3 through 6 we considered processes which involved discrete variables and also a discrete time-scale. We then proceeded, in Chapters 7 through 13, to discuss processes with discrete variables, but in continuous time. The next group, which we shall investigate in this chapter, is where we have continuous variables in continuous time. (There is of course a fourth classification for continuous variables in discrete time. This seems to be less important, and we shall not deal with this type of process.)

The discrete Markov processes in continuous time are characterized by the chance of a transition in a small interval of time being small, but the size of the transition when it occurs being appreciable. For example, in a simple birth process the probability of a birth in Δt in a population of size n is the small quantity $\lambda n \, \Delta t$, but when a birth does occur it entails the addition of one whole unit to the population, and this will be relatively large if n is small. On the other hand, in the continuous processes that we are now going to study it is certain that some change will occur in Δt, but for small Δt the change will also be small. We shall call these *diffusion processes*.

Many problems of biological importance involve relatively large populations subject to the transitions resulting from birth, death, mutation, infection, etc. When the populations are large, the transitions are relatively very small. They may also be relatively frequent for a suitably chosen time-scale. Under these conditions it may be possible to use an approximate model of the diffusion type in which both the variable and the time are continuous. The classical use of such models is of course in physics where we have a large number of particles executing a Brownian motion with small, rapid, random steps, the net result being to produce a diffusion effect on a macroscopic scale.

It is important to realize that this diffusion approach to large populations does not mean that we are adopting a deterministic treatment. The incorporation of small but rapid transitions into the model is done in such a way that an appreciable amount of statistical variation remains. An analogous situation occurs in the Central Limit Theorem, where we consider the cumulative effect of a large number of small chance elements. In the two following sections we shall see how a diffusion model can be obtained, by the appropriate limiting procedure, for two kinds of discrete processes. We shall then consider some general theory, and finally look at one or two specific applications.

14.2 Diffusion limit of a random walk

As we have seen, random-walk representations may provide another avenue of approach to the solution of the equations characterizing certain Markov processes. Of course, the random walk may itself be difficult to handle, as in the application to the total size of a general stochastic epidemic (see Section 12.3). There may therefore be some advantage in investigating the diffusion limit of a random-walk model. Let us see in a specific, elementary case how this is done.

We shall consider a simple, unrestricted random walk on the real integers, starting at the origin $x = 0$, and with probabilities p and $q = 1 - p$ of a transition at any stage of one unit to the right or left, respectively. Let u_{rn} be the probability that the nth step carries the particle to the point $x = r$. In the Markov chain notation of Chapter 5, the absolute probability distribution at the nth stage is thus given by $p_r^{(n)} = u_{rn}$. By using the kind of arguments already employed in Chapter 4 to obtain difference equations, we can easily see that

$$u_{r,n+1} = pu_{r-1,n} + qu_{r+1,n}, \tag{14.1}$$

with boundary conditions

$$u_{00} = 1; \qquad u_{r0} = 0, \quad r \neq 0. \tag{14.2}$$

Actually we can derive u_{rn} quite easily by a more direct method than solving (14.1). For if j of the n steps have been to the right and $n - j$ to the left, where

$$r = j - (n - j) = 2j - n, \tag{14.3}$$

the probability u_{rn} must be the binomial quantity

$$u_{rn} = \binom{n}{j} p^j q^{n-j} = \binom{n}{\frac{1}{2}n + \frac{1}{2}r} p^{\frac{1}{2}n + \frac{1}{2}r} q^{\frac{1}{2}n - \frac{1}{2}r}, \tag{14.4}$$

where only integral values of $\frac{1}{2}(n + r)$ are admissible.

Let us now consider the possibility of a limiting representation of this process. Suppose the units of distance are chosen so that the length of each step is Δx, and the motion of the particle is such that the time taken between any two consecutive positions is Δt. Thus during a period of time t, the particle makes a total of approximately $t/\Delta t$ steps, while the net displacement is $x/\Delta x$ steps. Before going to the limit we have a situation in which the time and distance can be only multiples of Δt and Δx, respectively. But in the limit, they can vary continuously.

There is a good deal of choice as to the actual mode of passing to the limit. We want a mode which has a useful and meaningful interpretation. Some discussion of possible alternatives is therefore necessary before making a decision. Now the maximum possible displacement in time t is $t\,\Delta x/\Delta t$. So that if we allow $\Delta x/\Delta t$ to tend to zero, no motion will be possible in the limiting situation. An appropriate ratio must be maintained if the process is not to degenerate.

During time t the total displacement is the sum of approximately $t/\Delta t$ independent random variables, each taking the values Δx and $-\Delta x$ with probabilities p and q, respectively, and therefore having mean $(p - q)\,\Delta x$ and variance $4pq(\Delta x)^2$. The mean and variance of the net displacement in time t are therefore approximately $t(p - q)\,\Delta x/\Delta t$ and $4pqt(\Delta x)^2/\Delta t$, respectively. If the limiting variance is to remain finite we must have $(\Delta x)^2/\Delta t = O(1)$. It then follows that, for the mean to stay finite, we require $p - q = O(\Delta x)$. These conditions will be satisfied if we put

$$\frac{(\Delta x)^2}{\Delta t} = 2D, \qquad p = \tfrac{1}{2} + \frac{c}{2D}\,\Delta x, \qquad q = \tfrac{1}{2} - \frac{c}{2D}\,\Delta x, \qquad (14.5)$$

where c and D are suitable constants. As defined by (14.5) they are the *drift* and the *diffusion coefficient*, respectively. The mean and variance of the net displacement at time t are now given by $2ct$ and $2Dt$, respectively. If $c = 0$, we have a symmetric random walk. But notice that in the limit, even with $c \neq 0$, we must have $p \to \tfrac{1}{2}$. Otherwise the particle would drift so quickly that there would be zero chance of finite displacements.

Let us therefore adopt a limiting procedure in which $\Delta x \to 0$ and $\Delta t \to 0$, subject to the conditions in (14.5). If n steps have occurred by about time t, we have $t \doteq n\,\Delta t$. As $\Delta t \to 0$, $n \to \infty$. And as $n \to \infty$, the binomial distribution in (14.4) will tend to a normal distribution with mean $2ct$ and variance $2Dt$.

For the units of distance and time chosen for carrying out the limiting procedure described above, the difference equation in (14.1) takes the form

$$u_{x,t+\Delta t} = pu_{x-\Delta x,t} + qu_{x+\Delta x,t}. \qquad (14.6)$$

Expanding by means of Taylor's theorem gives

$$\frac{\partial u}{\partial t}\Delta t = \frac{\partial u}{\partial x}(q - p)\Delta x + \frac{\partial^2 u}{\partial x^2}\frac{(\Delta x)^2}{2!}, \tag{14.7}$$

neglecting terms on the left of order $(\Delta t)^2$, and on the right of order $(\Delta x)^3$. Let us write $f(x, t)$ for the limiting form of $u_{x,t}$ as Δx, $\Delta t \to 0$, and we obtain a continuous frequency function. It is clear that the limiting form of (14.7) is the differential equation for the frequency function $f(x, t)$ given by

$$\frac{\partial f}{\partial t} = -2c\frac{\partial f}{\partial x} + D\frac{\partial^2 f}{\partial x^2}, \tag{14.8}$$

where we have made use of the conditions in (14.5). This is the well-known Fokker–Planck equation for diffusion with drift. In the usual theory the normal distribution is obtained analytically as a solution of (14.8).

Although the above derivation is only heuristic in nature it can provide a useful lead in discovering approximate and asymptotic solutions for discrete processes, by replacing the latter by suitable continuous-variable processes, which may turn out to be more tractable.

14.3 Diffusion limit of a discrete branching process

Let us now apply the above ideas to examine the limiting form of the discrete branching processes dealt with in Chapter 6. We have already obtained in (6.10) the equation

$$K_{n+1}(\theta) = K_n(K(\theta)), \tag{14.9}$$

where $K_n(\theta)$ is the cumulant-generating function for the population size X_n at the nth generation, and $K(\theta)$ is the cumulant-generating function for the number of offspring X produced by a single individual. We now change the scales of both the time and the random variable, writing

$$Y(t) = X_n\delta, \quad t = n\delta. \tag{14.10}$$

The cumulant-generating function of $Y(t)$, which we shall write as $K_y(\theta, t)$, can thus be expressed in the form

$$K_y(\theta, t) \equiv K_n(\theta\delta), \tag{14.11}$$

where the argument on the right is $\theta\delta$ instead of θ because of the change in scale (using a standard and easily proved result).

If we consider the cumulant-generating function for Y at time $t + \Delta t = (n + 1)\delta$, with $\Delta t = \delta$, we obtain from (14.9)

$$K_y(\theta, t + \Delta t) \equiv K_{n+1}(\theta\delta)$$

$$= K_n(K(\theta\delta))$$

$$= K_y(\delta^{-1}K(\theta\delta), t), \tag{14.12}$$

the last line following from (14.11) with $K(\theta\delta)$ substituted for $\theta\delta$.

We next examine the form of the function $K(\theta)$. If the distribution of the number of offspring of a single individual has mean m and variance σ^2,

$$K(\theta) = m\theta + \tfrac{1}{2}\sigma^2\theta^2 + \cdots . \tag{14.13}$$

It is reasonable to assume that the mean m can be written in the form

$$m = 1 + \mu\delta, \tag{14.14}$$

since as $\delta \to 0$ we shall want $m \to 1$, otherwise there would be a finite average increase in population size in an infinitesimal increment of time. Using the value of m given by (14.14), we have

$$\delta^{-1}K(\theta\delta) = \theta + (\mu\theta + \tfrac{1}{2}\sigma^2\theta^2)\delta + O(\delta^2). \tag{14.15}$$

If we now substitute (14.15) into the right-hand side of (14.12), we can expand about the value $K_y(\theta, t)$ by Taylor's theorem to give

$$K_y(\theta, t + \Delta t) = K_y(\delta^{-1}K(\theta\delta), t)$$

$$= K_y(\theta, t) + (\mu\theta + \tfrac{1}{2}\sigma^2\theta^2)\delta \frac{\partial K}{\partial \theta} + \cdots . \tag{14.16}$$

Since $\Delta t = \delta$, we can obtain immediately from (14.16)

$$\frac{\partial K}{\partial t} = \lim_{\Delta t \to 0} \frac{K_y(\theta, t + \Delta t) - K_y(\theta, t)}{\Delta t}$$

$$= (\mu\theta + \tfrac{1}{2}\sigma^2\theta^2) \frac{\partial K}{\partial \theta}. \tag{14.17}$$

Using $K = \log M$, the corresponding equation for the moment-generating function is

$$\frac{\partial M}{\partial t} = (\mu\theta + \tfrac{1}{2}\sigma^2\theta^2) \frac{\partial M}{\partial \theta}. \tag{14.18}$$

Finally, if $f(x, t)$ is the frequency function corresponding to $M(\theta, t)$, we have the standard inversion formula

$$f(x, t) = \frac{1}{2\pi} \int_{-\infty}^{\infty} e^{-\theta x} M(\theta, t) \, d\theta. \qquad (14.19)$$

Multiplying both sides of (14.18) by $(1/2\pi)e^{-\theta x}$, integrating with respect to θ, and using (14.19), yields

$$\frac{\partial f}{\partial t} = -\mu \frac{\partial (xf)}{\partial x} + \tfrac{1}{2}\sigma^2 \frac{\partial^2 (xf)}{\partial x^2}, \qquad (14.20)$$

with initial condition

$$f(0, t) = 0. \qquad (14.21)$$

Equation (14.20) is a special case of a forward Kolmogorov diffusion equation with infinitesimal mean and variance proportional to the population size. The general theory of models leading to this type of diffusion equation will be discussed in the following section. For the time being, we merely note that, under suitable limiting conditions, a discrete branching process can be replaced by a continuous-variable process in continuous time, whose frequency function satisfies a certain kind of partial differential equation.

14.4 General theory

In handling discrete variables in continuous time we found it convenient to define $p_{ij}(s, t)$ as the probability of finding the system in state E_j at time t, *given* that it was previously in state E_i at time s. Now that we are concerned with a continuous distribution of probability, a slightly different notation is required. Suppose that the random variable X representing the process in question takes the value y at time s, and x at time t, i.e. $X(s) = y$ and $X(t) = x$. Let us write $f(y, s; x, t)$ for the conditional frequency function of the variable x at time t, *given* the value y at the previous time s. Notice that the order of the pairs (y, s) and (x, t) represents the direction of the transition.

We now consider three epochs of time $s < t < u$, the corresponding variable-values being y, x, and z. The obvious analog of the Chapman–Kolmogorov equations for the discrete-variable case given in (7.41) is

$$f(y, s; z, u) = \int_{-\infty}^{\infty} f(y, s; x, t) \, f(x, t; z, u) \, dx, \qquad (14.22)$$

where we are assuming the usual Markov property that, given the present state of the system, the future behavior does not depend on the past. Equation (14.22) can easily be proved by considering first a path from (y, s) to (z, u) through a particular intermediate point (x, t). The probability of this specific path for a Markov process is $f(y, s; x, t) f(x, t; z, u)$. Thus the total probability for a transition from (y, s) to (z, u) is obtained by integrating over all possible intermediate points, i.e. integrating with respect to x, as shown in (14.22).

There are also continuous-time analogs of the Kolmogorov forward and backward equations, previously derived for discrete variables and exhibited in (7.51) and (7.53). Suppose we consider, for $s < t$, the difference

$$f(y, s - \Delta s; x, t) - f(y, s; x, t). \tag{14.23}$$

Using (14.22), we have, with appropriate changes in the notation,

$$f(y, s - \Delta s; x, t) = \int_{-\infty}^{\infty} f(y, s - \Delta s; z, s) f(z, s; x, t) \, dz. \tag{14.24}$$

Again, the integral over all possible transitions from any point must equal unity. Hence we can write

$$f(y, s; x, t) = f(y, s; x, t) \int_{-\infty}^{\infty} f(y, s - \Delta s; z, s) \, dz. \tag{14.25}$$

Combining (14.24) and (14.25) means that (14.23) can be written as

$(fy, s - \Delta s; x, t) - f(y, s; x, t)$

$$= \int_{-\infty}^{\infty} f(y, s - \Delta s; z, s)\{f(z, s; x, t) - f(y, s; x, t)\} \, dz$$

$$= \int_{-\infty}^{\infty} f(y, s - \Delta s; z, s)\left\{(z - y) \frac{\partial f(y, s; x, t)}{\partial y}\right.$$

$$\left. + \tfrac{1}{2}(z - y)^2 \frac{\partial^2 f(y, s; x, t)}{\partial y^2} + \cdots\right\} dz$$

$$\doteq \frac{\partial f(y, s; x, t)}{\partial y} \int_{-\infty}^{\infty} (z - y) f(y, s - \Delta s; z, s) \, dz$$

$$+ \frac{1}{2} \frac{\partial^2 f(y, s; x, t)}{\partial y^2} \int_{-\infty}^{\infty} (z - y)^2 f(y, s - \Delta s; z, s) \, dz. \tag{14.26}$$

Let us now suppose that there exist infinitesimal means and variances for changes in the basic random variable $X(t)$ defined by

$$
\left.
\begin{aligned}
a(y, s) &= \lim_{\Delta s \to 0} \frac{1}{\Delta s} \int (z - y) f(y, s - \Delta s; z, s)\, dz \\[1em]
b(y, s) &= \lim_{\Delta s \to 0} \frac{1}{\Delta s} \int (z - y)^2 f(y, s - \Delta s; z, s)\, dz
\end{aligned}
\right\}, \qquad (14.27)
$$

respectively. We may have to restrict the ranges of integration on the right of (14.27) to ensure convergence, but it is unnecessary to pursue this point in the present non-rigorous discussion.

Next, dividing both sides of (14.26) by Δs, letting $\Delta s \to 0$, and then using (14.27), gives

$$
-\frac{\partial f(y, s; x, t)}{\partial s} = a(y, s) \frac{\partial f(y, s; x, t)}{\partial y} + \tfrac{1}{2} b(y, s) \frac{\partial^2 f(y, s; x, t)}{\partial y^2}. \quad (14.28)
$$

This is the *backward Kolmogorov diffusion equation* for the continuous-variable, continuous-time, stochastic process introduced at the beginning of the section. A similar argument based on considering a forward increment in t, i.e. from t to $t + \Delta t$, instead of a backward increment in s as above, leads to the *forward Kolmogorov diffusion equation* (also called the *Fokker–Planck equation*) given by

$$
\frac{\partial f(y, s; x, t)}{\partial t} = -\frac{\partial}{\partial x} \{a(x, t) f(y, s; x, t)\} + \frac{1}{2} \frac{\partial^2}{\partial x^2} \{b(x, t) f(y, s; x, t)\}.
$$

$$(14.29)$$

As a rule, in physical applications the infinitesimal variance $b(x, t)$ is essentially positive. Sometimes, however, $b(x, t)$ vanishes at one, or both, of the boundaries. Again, one of the coefficients in the diffusion equation may have no finite limit at the boundaries. Such equations are said to be *singular diffusion equations*. This occurs typically in genetics, for example, with diffusion models involving gene frequencies (see Moran, 1962). Satisfactory treatment of the singular situation usually requires a special form of analysis, on account of the possibility that masses of probability will build up at the boundaries (see Bharucha–Reid, 1960, Sections 3.3 and 4.5).

A simple case of equation (14.29) was obtained in Section 14.3, where we considered the limiting form of a discrete branching process. The diffusion equation in (14.20) is clearly equivalent to the adoption of a continuous-variable model in which the infinitesimal mean and variance are proportional to the population size and are μx and $\sigma^2 x$, respectively.

It will be realized that, although we obtained diffusion processes in Section 14.2 and 14.3 as the limiting forms of certain kinds of discrete processes, the treatment of the present section actually starts with a model in which both the random variables and the time are continuous. This means that the theory of diffusion models can be developed in its own right, and not merely as an approximation to some other formulation.

Now, as usual, we call the process *homogeneous* if the transition probabilities depend only on the time difference $t - s$ (apart from x and y). Again, if the transition probabilities depend only on the increment $x - y$ (apart from s and t) we say that the process is *additive*. Suppose that, in the present diffusion type of model, we consider processes that are both homogeneous and additive. Then we can write

$$f(y, s; x, t) \equiv g(x - y, t - s), \qquad (14.30)$$

and, in addition, the infinitesimal means and variances will have to be constant, i.e.

$$a(x, t) \equiv a, \, b(x, t) \equiv b. \qquad (14.31)$$

Substitution of these forms in (14.28) and (14.29) shows that in this case the backward and forward equations can be reduced to the same partial differential equation, namely

$$\frac{\partial g(x, t)}{\partial t} = -a \frac{\partial g(x, t)}{\partial x} + \tfrac{1}{2} b \frac{\partial^2 g(x, t)}{\partial x^2}. \qquad (14.32)$$

Solution of the general diffusion equation

A satisfactory elucidation of the properties of a diffusion process depends to a great extent on our being able to handle the basic partial differential equations. Typically, we have the forward Kolmogorov equation, given in (14.29), which we shall write more compactly as

$$\frac{\partial f}{\partial t} = -\frac{\partial(af)}{\partial x} + \frac{1}{2} \frac{\partial^2(bf)}{\partial x^2}, \qquad (14.33)$$

together with appropriate boundary conditions. In general, there may be considerable difficulty in solving this second-order partial differential equation.

Let us now suppose that the process is homogeneous, i.e.

$$a(x, t) \equiv a(x), \qquad b(x, t) \equiv b(x). \qquad (14.34)$$

Using primes to indicate differentiation with respect to x, we may write (14.33) as

$$\frac{\partial f}{\partial t} = (\tfrac{1}{2} b'' - a') f + (b' - a') f' + \tfrac{1}{2} b f''. \qquad (14.35)$$

Suppose we look for a solution based on separating the variables. That is, write

$$f(x, t) = X(x)T(t),$$ (14.36)

and substitute in (14.35). (Note, of course, that $X(x)$ is not to be confused with the random-variable notation.) We obtain

$$X \frac{dT}{dt} = \tfrac{1}{2}bTX'' + (b' - a')TX' + (\tfrac{1}{2}b'' - a')TX,$$

or $$\frac{1}{T}\frac{dT}{dt} = X^{-1}\{\tfrac{1}{2}bX'' + (b' - a')X' + (\tfrac{1}{2}b'' - a')X\}.$$ (14.37)

It can be seen that the left of (14.37) involves only t, and the right only x. Thus we can equate both sides to the same constant $-\lambda$, say. We then have two ordinary differential equations

$$\frac{dT}{dt} = -\lambda T,$$ (14.38)

and $$\tfrac{1}{2}bX'' + (b' - a')X' + (\tfrac{1}{2}b'' - a' + \lambda)X = 0.$$ (14.39)

For any value of λ, say λ_r, we have $T = e^{-\lambda_r t}$ from (14.38). And we may hope to solve (14.39) by standard methods to yield a solution $X(x, \lambda_r) \equiv X_r$. The general solution of (14.35) can thus be expressed in the form

$$f(x, t) = \sum_{r=0}^{\infty} A_r X_r e^{-\lambda_r t},$$ (14.40)

where the λ_r and the A_r are determined by the boundary conditions. In general, the boundary conditions will indicate what values of λ_r are admissible: these are the eigenvalues of the solution of (14.39) and the corresponding $X(x, \lambda_r)$ are the eigenfunctions. For a more complete discussion of these methods see, for example, Sagan (1961).

An alternative approach is to use the Laplace-transform method, which we have employed in the discrete theory of queues and epidemics. Thus, taking transforms (defined in (11.49), for example) of both sides of (14.35), yields

$$\tfrac{1}{2}b \frac{d^2f^*}{dx^2} + (b' - a)\frac{df^*}{dx} + (\tfrac{1}{2}b'' - a' - s)f^* = -f(0).$$ (14.41)

This is a non-homogeneous ordinary differential equation of second order. The initial condition for $t = 0$ is automatically incorporated, but in attempting to derive a solution we must impose the transformed version of any other boundary conditions that may be required.

In the special case of additive processes which satisfy an equation of the form in (14.32), solutions may be obtained fairly easily. By comparison with (14.8), which we derived in a somewhat heuristic manner as the limiting form of a random walk, it is clear that a particular solution of (14.32) must be a normal distribution of the form

$$g(x, t) = \frac{1}{(2\pi bt)^{\frac{1}{2}}} \exp\left\{-\frac{(x - at)^2}{2bt}\right\}. \tag{14.42}$$

This solution corresponds to the diffusion limit for a random walk starting at the origin $x = 0$ when $t = 0$. A general solution of (14.32) can then be obtained by a linear superposition of particular solutions starting from arbitrary points.

Suppose, for instance, that there are absorbing barriers at $x = \pm d$ ($d > 0$). Then it can be shown that the general solution for $a = 0$ is

$$g(x, t) = \sum_{j=-\infty}^{\infty} (-)^j (2\pi bt)^{-\frac{1}{2}} \exp\{-(x - 2d)^2/2bt\}. \tag{14.43}$$

The case in which the left-hand barrier tends to $x = -\infty$, i.e. in effect a single absorbing barrier at $x = d$, has the solution

$$g(x, t) = (2\pi bt)^{-\frac{1}{2}} \left\{\exp\left(-\frac{x^2}{2bt}\right) - \exp\left(-\frac{(x - 2d)^2}{2bt}\right)\right\}. \tag{14.44}$$

More generally, if $a \neq 0$, we multiply $g(x, t)$ in (14.43) and (14.44) by a factor $\exp\{(ax - \frac{1}{2}a^2 t)/b\}$.

The distribution of the *passage-time* to the boundaries can be obtained by noting that, as time proceeds, part of the total probability is lost at the boundaries. Thus if the frequency function of the passage-time to the boundaries is $h(t)$, we have

$$h(t) = -\frac{\partial}{\partial t} \int_{-d_1}^{d_2} g(x, t) \, dx, \tag{14.45}$$

for boundaries at $x = -d_1$ and d_2. And in the special case when $d_1 = \infty$ and $d_2 = d$, we find that

$$h(t) = \frac{d}{(2\pi bt^3)^{\frac{1}{2}}} \exp\left\{-\frac{(d - at)^2}{2bt}\right\}. \tag{14.46}$$

For further details of the difficulties that arise in trying to obtain solutions of diffusion equations (especially when those are of the so-called "singular" type) see Chapters 4 and 5 of Bharucha–Reid (1960), which also include useful bibliographies.

14.5 Application to population growth

Let us now examine an application of the diffusion type of approach to a stochastic population model involving both birth and death. For sufficiently large population sizes we could adopt a model for which the forward Kolmogorov equation was given by (14.33), i.e.

$$\frac{\partial f}{\partial t} = -\frac{\partial (af)}{\partial x} + \frac{1}{2}\frac{\partial^2 (bf)}{\partial x^2},$$

where the infinitesimal mean a and variance b must be suitably chosen. It seems reasonable to assume that the latter quantities are in fact proportional to the population size. We can then write

$$a(x) = \alpha x, \qquad b(x) = \beta x,$$

where $\beta > 0$, and the drift α is positive or negative as the expected increase in population size is positive or negative. The appropriate diffusion equation is thus

$$\frac{\partial f}{\partial t} = -\alpha \frac{\partial (xf)}{\partial x} + \frac{1}{2}\beta \frac{\partial^2 (xf)}{\partial x^2}. \qquad (14.48)$$

If the initial condition is $x = x_0$, $t = 0$, the solution of (14.48) can be shown to be

$$f(x, t) = \frac{2\alpha}{\beta(e^{\alpha t} - 1)}\left(\frac{x_0 e^{\alpha t}}{x}\right)^{\frac{1}{2}}\exp\left\{\frac{-2\alpha(x_0 e^{\alpha t} + x)}{\beta(e^{\alpha t} - 1)}\right\}I_1\left(\frac{4\alpha(x_0 x e^{\alpha t})^{\frac{1}{2}}}{\beta(e^{\alpha t} - 1)}\right), \qquad (14.49)$$

where I_1 is a Bessel function of first order.

The stochastic mean $m(t)$ can be obtained either in the usual way from (14.49), or more easily from the basic partial differential equation (14.48) as follows.

We have

$$\frac{dm(t)}{dt} = \frac{d}{dt}\int_0^\infty xf \, dx$$

$$= \int_0^\infty x\frac{\partial f}{\partial t} \, dx$$

$$= -\alpha \int_0^\infty x\frac{\partial (xf)}{\partial x} \, dx + \frac{1}{2}\beta \int_0^\infty x\frac{\partial^2 (xf)}{\partial x^2} \, dx, \quad \text{using (14.48)},$$

$$= \alpha m(t), \qquad (1.450)$$

provided $x^2 f$ and $x^2 \partial f / \partial x$ both vanish at the end-points of the integration.

From (14.50) we derive by integration the result

$$m(t) = x_0 e^{\alpha t}, \tag{14.51}$$

which shows that the stochastic mean follows the familiar exponential rate of growth with parameter α. In a similar way we can obtain the variance as

$$\sigma^2(t) = (\beta/\alpha)x_0 e^{\alpha t}(e^{\alpha t} - 1). \tag{14.52}$$

(Note that σ^2 is still positive for $\alpha < 0$, since then $e^{\alpha t} < 1$.)

It will be realized that even if $\alpha > 0$ the type of assumptions involved in a diffusion model implies the possibility of losses as well as gains in population, i.e. we have a model that incorporates both birth and death. It is natural, therefore, to inquire about the probability of extinction. Now the total frequency given by integrating $f(x, t)$ in (14.49) from 0 to ∞ is

$$\int_0^\infty f(x, t)\, dt = 1 - \exp\left\{\frac{-2\alpha x_0 e^{\alpha t}}{\beta(e^{\alpha t} - 1)}\right\}. \tag{14.53}$$

This quantity is less than unity, the balance of probability, given by

$$p_0(t) = \exp\left\{\frac{-2\alpha x_0 e^{\alpha t}}{\beta(e^{\alpha t} - 1)}\right\}, \tag{14.54}$$

corresponding to the frequency absorbed at $x = 0$ by time t, i.e. the proportion of processes which will have become extinct by time t. The limiting values of the extinction probability are

$$\left.\begin{aligned} \lim_{t \to \infty} p_0(t) &= 1, \qquad \alpha \geqslant 0 \\ &= e^{-2\alpha x_0/\beta}, \qquad \alpha > 0 \end{aligned}\right\}. \tag{14.55}$$

It is interesting to compare these results with the continuous-time birth-and-death process of Section 8.6, for which the birth- and death-rates were λ and μ, respectively. Suppose we look for a suitable continuous-variable diffusion analog. Now the transitions which may occur in Δt are $+1$ with probability $\lambda x\, \Delta t$ and -1 with probability $\mu x\, \Delta t$. The mean and variance of this latter distribution are easily found to be $(\lambda - \mu)x\, \Delta t$ and $(\lambda + \mu)x\, \Delta t$, to first order in Δt. We should thus choose the infinitesimal mean and variance as $(\lambda - \mu)x$ and $(\lambda + \mu)x$, i.e. $\alpha = \lambda - \mu$ and $\beta = \lambda + \mu$.

The result in (14.51) thus corresponds exactly to (8.84) with $x_0 = a$, and (14.52) corresponds to (8.49). There is also considerable similarity between the forms of the extinction probabilities given in (14.55) and (8.59). The value unity is obtained in both cases when $\lambda - \mu = \alpha \leqslant 0$; although the precise expressions differ when $\lambda - \mu = \alpha > 0$.

CHAPTER 15

Approximations to Stochastic Processes

15.1 Introduction

It is clear from the discussions of stochastic processes undertaken so far in this book, that even the restriction to Markov chains and processes does not prevent many conceptually quite simple models from entailing heavy and relatively intractable mathematics. A number of difficult situations arose when second-order partial differential equations for generating functions were involved, as with epidemics, competition, and predation. It is likely that further mathematical advances will overcome at least some of these difficulties. At the same time it is worth considering the feasibility of various types of approximation—always a useful approach in applied mathematics. The present chapter is in no sense a complete introduction to this topic. But it does contain a number of devices, some based on analytical approximations, some derived from stochastic modifications of the basic model, that may well be useful in practice. This whole approach has so far received comparatively little systematic attention, but seems likely to become of increasing importance in the future.

15.2 Continuous approximations to discrete processes

We have already seen in Chapter 14 how in some circumstances it is advantageous to work with a diffusion model that approximates to the basic discrete-variable process. In choosing the appropriate diffusion model we have to be careful and adopt the proper limiting procedure for the scales of both variable and time. A related approach that avoids some of these difficulties (discussed in detail by Daniels, 1960), is to seek a continuous solution to the basic differential-difference equations of a discrete-variable process. In effect, this simply means that we try to find a suitable continuous frequency curve passing approximately through the individual values of a discrete probability distribution.

The Poisson process

Let us consider first a Poisson process with parameter λ (see Section 7.2), for which the basic differential-difference equation for the probability $p_n(t)$ of having n individuals at time t is given by

$$\frac{dp_n}{dt} = \lambda p_{n-1} - \lambda p_n, \tag{15.1}$$

with initial condition

$$p_0(0) = 1. \tag{15.2}$$

The exact solution is of course

$$p_n(t) = \frac{e^{-\lambda t}(\lambda t)^n}{n!}, \quad n = 0, 1, 2, \cdots. \tag{15.3}$$

Since we are interested in the possibility of a continuous solution of (15.1), let us replace $p_n(t)$ by $p(x, t)$, where x is now continuous. Thus (15.1) becomes

$$\frac{\partial p(x, t)}{\partial t} = \lambda p(x - 1, t) - \lambda p(x, t). \tag{15.4}$$

If we now expand the first term on the right-hand side of (15.4) in a Taylor series about $p(x, t)$, we obtain

$$\frac{\partial p(x, t)}{\partial t} = \lambda \left\{ p(x, t) - \frac{\partial p(x, t)}{\partial x} + \frac{1}{2} \frac{\partial^2 p(x, t)}{\partial x^2} \cdots \right\} - \lambda p(x, t).$$

Retaining only terms of second order yields the equation

$$\frac{\partial p}{\partial t} = -\lambda \frac{\partial p}{\partial x} + \tfrac{1}{2}\lambda \frac{\partial^2 p}{\partial x^2}, \tag{15.5}$$

which is in fact of diffusion type, with solution

$$p(x, t) = \frac{1}{(2\pi\lambda t)^{\frac{1}{2}}} \exp\left\{ -\frac{(x - \lambda t)^2}{2\lambda t} \right\}. \tag{15.6}$$

For λt not too small, this normal distribution is the usual reasonably good approximation to the exact Poisson distribution in (15.3). The disadvantage of (15.6) is that it gives non-zero probabilities for negative values of x. Although these will be negligible for sufficiently large λt, they will be appreciable for small λt. Thus the *relative* errors of the approximation become important for small x.

One way of trying to overcome this last difficulty is to work with $\log p$ instead of p, since $\delta(\log p) = p^{-1}\delta p$. Suppose we write

$$l(x, t) = \log p(x, t). \tag{15.7}$$

Substituting from (15.7) into (15.4) yields

$$\frac{\partial l(x, t)}{\partial t} = e^{-l(x,t)}\{\lambda e^{l(x-1,t)} - \lambda e^{l(x,t)}\}$$

$$= \lambda\{e^{l(x-1,t)-l(x,t)} - 1\}. \tag{15.8}$$

This time let us expand the exponent on the right only as far as the first differential coefficient to give

$$\frac{\partial l}{\partial t} = \lambda\left\{\exp\left(-\frac{\partial l}{\partial x}\right) - 1\right\}. \tag{15.9}$$

The equation in (15.9) can be solved by standard methods (e.g. Forsyth, 1929, p. 412). We put

$$-\frac{\partial l}{\partial x} = a, \qquad \frac{\partial l}{\partial t} = \lambda(e^a - 1), \tag{15.10}$$

for which the complete integral is the two-parameter family of planes

$$l(x, t) = \lambda t(e^a - 1) - ax + C. \tag{15.11}$$

The solution we require is the one-parameter envelope of these planes, with $C = C(a)$, having the correct form at $t = 0$. The initial condition (15.2), in the present notation, means that we should like $p(x, 0) = 0$ if $x \neq 0$ and $p(0, 0) = 1$. We cannot expect the approximation to work near $t = 0$, but let us make the planes in (15.11) pass through the point $l = 0$, $t = 0$, $x = 0$. It follows that $C = 0$, and the envelope is then given by

$$\left.\begin{array}{l} l = \lambda t(e^a - 1) - ax \\ 0 = \lambda t e^a - x \end{array}\right\}. \tag{15.12}$$

Eliminating a from these equations thus provides the solution

$$l(x, t) = x - \lambda t - x \log(x/\lambda t),$$

or

$$p(x, t) = \frac{e^{-\lambda t}(\lambda t)^x}{x^x e^{-x}}. \tag{15.13}$$

Comparison of (15.13) with the exact solution (15.3) shows that the method works surprisingly well, in spite of the crude and heuristic nature of the argument and the retention of only first-order derivatives in (15.9). The approximation in (15.13) involves only positive values of x. If we put $x = n$, it is clear that the main difference between (15.3) and (15.13) is that the denominator in the former is replaced by a crude form of Stirling's approximation to $n!$. An improvement in the approximation in (15.13) is obtained if a suitable multiplying constant A is introduced. This means, in effect, using a more accurate version of Stirling's formula. We could obtain such a result by making the planes in (15.11) pass through the point $l = \log A$, $t = 0$, $x = 0$, in which case $C = \log A$.

The main objection to the type of analysis outlined above is that the required approximate solution is determined by the initial conditions when $t = 0$, and it is precisely at this point that the solution breaks down.

Birth processes

We next consider a general birth process in which the birth-rate at any time is λ_n, depending on the population size n (see Section 8.3). The basic differential-difference equation for the probabilities $p_n(t)$ is given by

$$\frac{dp_n}{dt} = \lambda_{n-1}p_{n-1} - \lambda_n p_n. \tag{15.14}$$

As we are looking for a continuous-variable approximation, we replace $p_n(t)$ by $p(x, t)$, and λ_n by $\lambda(x)$. We thus have

$$\frac{\partial p(x, t)}{\partial t} = \lambda(x - 1)p(x - 1, t) - \lambda(x)p(x, t). \tag{15.15}$$

Let us suppose that initially $x = \xi$ when $t = 0$.

If we again use the substitution in (15.7), we find that (15.15) becomes

$$\frac{\partial l(x, t)}{\partial t} = \lambda(x - 1)e^{l(x-1,t)-l(x,t)} - \lambda(x). \tag{15.16}$$

Let us now suppose that $\lambda(x)$ varies smoothly with x, and that $\lambda'(x)$ is small compared with $\lambda(x)$, at least for x not too small. Equation (15.16) can then be written approximately, as

$$\frac{\partial l}{\partial t} = \lambda(x)\left\{\exp\left(-\frac{\partial l}{\partial x}\right) - 1\right\}. \tag{15.17}$$

This is similar to (15.9) with $\lambda(x)$ instead of λ.

We can use the same method of solution as before. But if we put $-\partial l/\partial x = a$ we do not obtain an equation for $\partial l/\partial t$ independent of x. Accordingly, we write

$$\frac{\partial l}{\partial t} = a, \qquad \frac{\partial l}{\partial x} = -\log\left\{\frac{a}{\lambda(x)} + 1\right\}. \tag{15.18}$$

The complete integral of (15.18) is given by

$$l(x, t) = at + \int_{\xi}^{x} \log\left\{\frac{\lambda(w)}{a + \lambda(w)}\right\} dw + C. \qquad (15.19)$$

If we make the planes pass through the point $l = \log A$, $t = 0$, $x = 0$, we have $C = \log A$. The envelope with respect to a is thus obtained by eliminating a from the equations

$$p(x, t) = A \exp\left[at + \int_{\xi}^{x} \log\left\{\frac{\lambda(w)}{a + \lambda(w)}\right\} dw\right],$$

$$0 = t - \int_{\xi}^{x} \frac{dw}{a + \lambda(w)} \qquad (15.20)$$

where the first line of (15.20) is obtained simply by taking exponentials of (15.19).

In the simple birth process we have $\lambda(x) \equiv \lambda x$ (see Section 8.2). The second equation of (15.20) can then be integrated explicitly to give a as a function of t, namely

$$a = \frac{\lambda(x - \xi e^{\lambda t})}{e^{\lambda t} - 1}. \qquad (15.21)$$

Substitution of this value in the first line of (15.20) then yields

$$p(x, t) = \frac{Ax^x e^{-\xi \lambda t}(1 - e^{-\lambda t})^{x - \xi}}{\xi^{\xi}(x - \xi)^{x - \xi}}. \qquad (15.22)$$

If we compare (15.22) with the exact result in equation (8.15), with $a = \xi$ and $n = x$, we see that the effect of the approximation is again to replace the factorials in $\begin{pmatrix} x - 1 \\ \xi - 1 \end{pmatrix}$ by crude Stirling expressions.

Laplace transform approach

One major difficulty, that of obtaining approximations of a reasonably good relative accuracy, has been overcome by working with the logarithm of the frequency function, as given by (15.7). Another awkward aspect of the treatment is the problem of satisfying the initial conditions. Now if we used Laplace transforms with respect to time, the initial conditions would be incorporated automatically into the basic difference equation. Let us, therefore, combine both methods, and investigate the logarithm of the Laplace transform of the frequency function.

We use the transform as previously defined in (11.49), so that

$$p^*(x, s) = \int_{0}^{\infty} e^{-st} p(x, t) \, dt, \qquad (15.23)$$

with

$$p(\xi, 0) = 1. \tag{15.24}$$

Thus, taking transforms of both sides of (15.15), we obtain

$$\left.\begin{aligned} sp^*(x, s) &= \lambda(x - 1)p^*(x - 1, s) - \lambda(x)p^*(x, s), \quad x > \xi \\ \text{and } sp^*(\xi, s) - 1 &= -\lambda(\xi)p^*(\xi, s) \end{aligned}\right\}. \tag{15.25}$$

Now let us write

$$L(x, s) = \log p^*(x, s). \tag{15.26}$$

Equations (15.25) can then be expressed as

$$\left.\begin{aligned} se^{L(x,s)} &= \lambda(x - 1)e^{L(x-1,s)} - \lambda(x)e^{L(x,s)}, \quad x > \xi \\ se^{L(\xi,s)} - 1 &= -\lambda(\xi)e^{L(\xi,s)} \end{aligned}\right\}. \tag{15.27}$$

So far the equations are valid for discrete x. Next, we view x as a continuous variable, and expand $L(x - 1, s)$ about $L(x, s)$ in a Taylor's series. The first line of (15.27) can thus be put in the form

$$s \doteq \lambda(x)\{e^{L(x-1,s)-L(x,s)} - 1\}$$

$$\doteq \lambda(x)\left\{\exp\left(-\frac{\partial L}{\partial x}\right) - 1\right\}, \tag{15.28}$$

expanding only as far as the first differential coefficient, and assuming that $\lambda'(x)$ is small. Taking logarithms of (15.28) leads to

$$\frac{\partial L}{\partial x} = -\log\left\{1 + \frac{s}{\lambda(x)}\right\}, \tag{15.29}$$

which integrates to give a first approximation

$$L_1(x, s) = -\int_\xi^x \log\left\{1 + \frac{s}{\lambda(w)}\right\} dw - \log\{s + \lambda(\xi)\}, \tag{15.30}$$

satisfying the initial conditions $x \doteq \xi$, $t = 0$.

It is quite easy to proceed to a second approximation, taking the first neglected terms of (15.27) into account. Thus instead of (15.28) we have

$$s = \lambda(x - 1)e^{L(x-1,s)-L(x,s)} - \lambda(x),$$

or

$$s + \lambda(x) \doteq \{\lambda(x) - \lambda'(x)\}\exp\left\{-\frac{\partial L}{\partial x} + \frac{1}{2}\frac{\partial^2 L}{\partial x^2}\right\}.$$

If we now assume that $\lambda'(x)/\lambda(x)$ is small, dividing by $\lambda(x)$ and taking logarithms yields

$$\log\left\{1 + \frac{s}{\lambda(x)}\right\} = -\frac{\lambda'(x)}{\lambda(x)} - \frac{\partial L}{\partial x} + \frac{1}{2}\frac{\partial^2 L}{\partial x^2}. \tag{15.31}$$

In carrying through this type of successive approximation the usual procedure at the present point is to replace the relatively small, analytically awkward, term $\partial^2 L/\partial x^2$ in (15.31) by the value $\partial^2 L_1/\partial x^2$ derived from the first approximation in (15.30). Now $\partial L_1/\partial x$ is given by the right-hand side of (15.29), and differentiating again yields

$$\frac{\partial^2 L_1}{\partial x^2} = \frac{\lambda'(x)}{\lambda(x)} - \frac{\lambda'(x)}{s + \lambda(x)}. \tag{15.32}$$

Substituting (15.32) on the right of (15.31) leads, after a little rearrangement, to

$$\frac{\partial L}{\partial x} = -\log\left\{1 + \frac{s}{\lambda(x)}\right\} - \frac{\lambda'(x)}{2\lambda(x)} - \frac{\lambda'(x)}{2\{s + \lambda(x)\}}. \tag{15.33}$$

We can integrate (15.33) to provide a new solution given by

$$L_2(x, s) = -\int_\xi^x \log\left\{1 + \frac{s}{\lambda(w)}\right\} dw - \tfrac{1}{2}\log\lambda(x) - \tfrac{1}{2}\log\{s + \lambda(x)\} + C,$$

where C is determined by putting $x = \xi$, i.e.

$$C = L_2(\xi, s) + \tfrac{1}{2}\log\lambda(\xi) + \tfrac{1}{2}\log\{s + \lambda(\xi)\}$$
$$= \log p^*(\xi, s) + \tfrac{1}{2}\log\lambda(\xi) + \tfrac{1}{2}\log\{s + \lambda(\xi)\}$$
$$= \tfrac{1}{2}\log\lambda(\xi) - \tfrac{1}{2}\log\{s + \lambda(\xi)\},$$

using the value of $p^*(\xi, s)$ obtained from the second equation in (15.25). Thus we can write

$$L_2(x, s) = -\int_\xi^x \log\left\{1 + \frac{s}{\lambda(w)}\right\} dw + \tfrac{1}{2}\log\frac{\lambda(\xi)}{\lambda(x)\{s + \lambda(x)\}\{s + \lambda(\xi)\}}.$$
$$\tag{15.34}$$

The required approximation to $p(x, t)$ is obtained by applying the inverse Laplace transform to (15.34). Formally, we have

$$p_2(x, t) = \frac{1}{2\pi i}\int_{c-i\infty}^{c+i\infty}\left[\frac{\lambda(\xi)}{\lambda(x)\{s + \lambda(x)\}\{s + \lambda(\xi)\}}\right]^{\frac{1}{2}}$$
$$\times \exp\left[st - \int_\xi^x \log\left\{1 + \frac{s}{\lambda(w)}\right\} dw\right] ds. \tag{15.35}$$

Although the expression in (15.35) appears to be rather complicated, it is fortunate that a sufficiently good approximation of the saddle-point type can easily be obtained (see following section for a discussion of this technique). We have to look for a point s_0 where the exponentiated function is stationary with respect to s. Thus

$$t = \int_{\xi}^{x} \frac{dw}{s_0 + \lambda(w)}. \tag{15.36}$$

Formula (15.48) applied to (15.35) then gives

$$p_2(x, t) \doteq \frac{\exp\left[s_0 t - \int_{\xi}^{x} \log\left\{1 + \frac{s_0}{\lambda(w)}\right\} dw\right]}{\left[\frac{2\pi\lambda(x)\{s_0 + \lambda(x)\}\{s_0 + \lambda(\xi)\}}{\lambda(\xi)} \int_{\xi}^{x} \frac{dw}{\{s_0 + \lambda(w)\}^2}\right]^{\frac{1}{2}}}. \tag{15.37}$$

In the special case of a birth process, where $\lambda(x) \equiv \lambda x$, we find that $p_2(x, t)$ in (15.37) reduces after some algebra to

$$p(x, t) = \frac{x^{x-\frac{1}{2}} e^{-\xi\lambda t}(1 - e^{-\lambda t})^{x-\xi}}{(2\pi)^{\frac{1}{2}}\xi^{\xi-\frac{1}{2}}(x - \xi)^{x-\xi-\frac{1}{2}}}. \tag{15.38}$$

This is a much better approximation than (15.22), and comparison with (8.15) with $a = \xi$ and $n = x$ shows that this time we have achieved an approximation which differs from the exact result only in replacing the factorials by their precise Stirling equivalents.

Of course, with more general functions $\lambda(x)$ we might have greater difficulty in obtaining the appropriate saddle-point approximations. However, the success of the whole method in the foregoing application suggests that the Daniels' approach may prove to be an extremely valuable tool in the handling of relatively intractable discrete-variable continuous-time stochastic processes.

15.3   Saddle-point approximations

The discussion of the Laplace transform method of the last section, resulting in the somewhat complicated formula (15.35), turned out to be fairly tractable because we were able to use a saddle-point type of approximation. This is essentially the dominant term in some relevant asymptotic expansion obtained by the method of steepest descents. For a rigorous account of the latter, see Copson (1948, p. 330) or DeBruijn (1958, p. 60).

In the present section we merely consider the derivation of the saddle-point term in a heuristic way, as this technique can be extremely valuable in handling otherwise unmanageable expressions and seems not to be as widely appreciated as it should.

The simplest type of application involving the ideas used in the saddle-point method is to real integrals of the type

$$F(t) = \int_{-\infty}^{\infty} \phi(x, t) \, dx, \qquad (15.39)$$

where the parameter t is taken to be large, and the function $\phi(x, t)$ has a fairly sharp peak somewhere so that the main contribution to the integral comes from the neighborhood of this peak.

Suppose we take, for example,

$$F(t) = \int_{-\infty}^{\infty} e^{-tx^2 \log(1 + x + x^2)} \, dx \qquad (15.40)$$

When t is large the integrand is everywhere very small except when x is near zero. Now

$$\log(1 + x + x^2) = \log \frac{1 - x^3}{1 - x} = x + \tfrac{1}{2}x^2 - \tfrac{2}{3}x^3 + \cdots,$$

where the range of x must be suitably restricted, say $-\tfrac{1}{2} < x < \tfrac{1}{2}$. It follows that

$$F(t) \doteq \int_{-\infty}^{\infty} e^{-tx^2}(x + \tfrac{1}{2}x^2 - \tfrac{2}{3}x^3) \, dx, \qquad (15.41)$$

since the main contributions to the integral come from the range within which the logarithmic expansion is valid. We can now integrate (15.41) quite easily to give the asymptotic result for large t

$$F(t) \sim \tfrac{1}{4}\pi^{\frac{1}{2}}t^{-\frac{3}{2}}. \qquad (15.42)$$

In general it is not necessary for the peak to be sharp, nor for its position to be independent of t: both of these can be adequately controlled by the use of a suitable substitution $x = a(t) + b(t)y$.

Next, let us consider integrals of the type

$$F(t) = \int_{-\infty}^{\infty} e^{tf(x)} \, dx, \qquad (15.43)$$

where $f(x)$ has an absolute maximum at $x = 0$. Expanding the exponent gives

$$F(t) = \int_{-\infty}^{\infty} \exp[t\{f(0) + xf'(0) + \tfrac{1}{2}x^2 f''(0) + \cdots\}]\, dx$$

$$= e^{tf(0)} \int_{-\infty}^{\infty} \exp\{\tfrac{1}{2}tf''(0)x^2 + \cdots\}\, dx$$

$$\sim \left\{\frac{2\pi}{-tf''(0)}\right\}^{\frac{1}{2}} e^{tf(0)}, \tag{15.44}$$

where we have used the fact that $f'(0) = 0$; note also that $f''(0) < 0$.

Now suppose we have an inverse Laplace transform integral of the type

$$F(t) = \frac{1}{2\pi i} \int_{c-i\infty}^{c+i\infty} e^{tf(s)} g(s)\, ds, \tag{15.45}$$

where again t may be thought of as large, though this may not be necessary. Note that (15.45) is a complex integral in the s-plane. It can be shown that if s_0 is a zero of $f'(s)$, i.e. $f'(s_0) = 0$, then the surface given by $w = \mathbf{R} f(s)$ has a saddle-point at $s = s_0$. The full method of steepest descents involves choosing a path of integration along a line of steepest descent from s_0 which can be obtained from the path in (15.45) by a suitable deformation. For such a path of integration the main contributions to the integral will come from the neighborhood of s_0.

Less precisely, we assume that there is a real point s_0, and put $c = s_0$ in (15.45). We then write $s = s_0 + i\sigma$. The integral then becomes

$$F(t) = \frac{1}{2\pi} \int_{-\infty}^{\infty} e^{tf(s_0 + i\sigma)} g(s_0 + i\sigma)\, d\sigma$$

$$= \frac{1}{2\pi} \int_{-\infty}^{\infty} \exp\{tf(s_0) - \tfrac{1}{2}tf''(s_0)\sigma^2 + \cdots\} g(s_0 + i\sigma)\, d\sigma,$$

$$\text{since} \quad f'(s_0) = 0,$$

$$\doteqdot \frac{e^{tf(s_0)}}{2\pi} \int_{-\infty}^{\infty} e^{-\frac{1}{2}tf''(s_0)\sigma^2} g(s_0 + i\sigma)\, d\sigma$$

$$\doteqdot \frac{e^{tf(s_0)}}{2\pi} \int_{-\infty}^{\infty} e^{-\frac{1}{2}tf''(s_0)\sigma^2} g(s_0)\, d\sigma$$

$$= \frac{g(s_0) e^{tf(s_0)}}{\{2\pi t f''(s_0)\}^{\frac{1}{2}}}. \tag{15.46}$$

A similar treatment of

$$F(t) = \frac{1}{2\pi i} \int_{c-i\infty}^{c+i\infty} e^{st - f(s)} g(s)\, ds \tag{15.47}$$

leads to the asymptotic form

$$F(t) \sim \frac{g(s_0)e^{s_0 t - f(s_0)}}{\{-2\pi f''(s_0)\}^{\frac{1}{2}}}.$$ (15.48)

This formula has been used above in approximating the inverse transform in (15.35).

15.4 Neglect of high-order cumulants

In many of the simpler birth-and-death type of processes it is possible to derive explicit expressions for the probability-generating function. When this attempt fails we can usually still fall back on the method of equating coefficients of powers of θ in the partial differential equation for the moment-generating function (or cumulant-generating function). If the transition probabilities are only linear functions of the random variables involved we usually obtain a set of ordinary differential equations for the moments (or cumulants), which can be solved successively (see, for example, the end of Section 8.6).

If, on the other hand, the transition probabilities are non-linear functions of the random variables, we are likely to find that the first equation involves the second moment as well as the first. We then have no basis for successive solution. This occurs, for example, with epidemics. Consider the simple stochastic epidemic of Section 12.2, for which the moment-generating function of the number of susceptibles at time t satisfies the partial differential equation previously given in (12.7), namely

$$\frac{\partial M}{\partial \tau} = (e^{-\theta} - 1)\left\{(n + 1)\frac{\partial M}{\partial \theta} - \frac{\partial^2 M}{\partial \theta^2}\right\},$$ (15.49)

with initial condition

$$M(\theta, 0) = e^{n\theta},$$ (15.50)

where the time-scale has been chosen as $\tau = \beta t$ (for an infection-rate β).

Equating coefficients of powers of θ on both sides of (15.49) leads to the equations (12.16), of which the first is

$$\frac{d\mu_1'}{d\tau} = -(n + 1)\mu_1' + \mu_2',$$ (15.51)

where μ_k' is the kth moment about the origin. Although it is possible to obtain μ_1' by alternative means (leading to the formula in (12.18)), the present approach does not yield a set of equations which can be solved successively: the first member of the set given by (15.51) contains μ_2' as well as μ_1'.

Suppose, however, that we work with the cumulant-generating function We put $K(\theta, t) = \log M(\theta, t)$. Substitution in (15.49) then gives

$$\frac{\partial K}{\partial \tau} = (e^{-\theta} - 1)\left\{(n + 1)\frac{\partial K}{\partial \theta} - \left(\frac{\partial K}{\partial \theta}\right)^2 - \frac{\partial^2 K}{\partial \theta^2}\right\}, \qquad (15.52)$$

a non-linear equation which looks even less promising. Let us assume that, for sufficiently large n, we could neglect all cumulants of order higher than j, say. The set of j differential equations obtained by equating coefficients of powers of θ up to θ^j on both sides of (15.52) then forms a closed system of j equations in j unknowns. This is, at least in principle, soluble. The first two equations are easily found to be

$$\frac{d\kappa_1}{d\tau} = -(n + 1)\kappa_1 + \kappa_1{}^2 + \kappa_2, \qquad (15.53)$$

and $$\frac{d\kappa_2}{d\tau} = (n + 1)\kappa_1 - \kappa_1{}^2 - (2n + 3)\kappa_2 + 4\kappa_1\kappa_2 + 2\kappa_3. \quad (15.54)$$

The first equation (15.53) is of course identical with (15.51) except for a change of notation, i.e. $\kappa_1 \equiv \mu_1'$ and $\kappa_2 \equiv \mu_2' - \mu_1'^2$. If now we could assume that, for sufficiently large n, all cumulants higher than the second could be neglected, i.e. that the distribution of susceptibles at any time was approximately normal, we could set $\kappa_3 = 0$ in (15.54) to give

$$\frac{d\kappa_2}{d\tau} \doteqdot (n + 1)\kappa_1 - \kappa_1{}^2 - (2n + 3)\kappa_2 + 4\kappa_1\kappa_2. \qquad (15.55)$$

Equations (15.53) and (15.55) thus involve only κ_1 and κ_2. It is not difficult to eliminate κ_2 by differentiation of (15.53) and substitution. The resulting equation for κ_1 is

$$\frac{d^2\kappa_1}{d\tau^2} + (3n + 4 - 6\kappa_1)\frac{d\kappa_1}{d\tau} + 2\kappa_1(n + 1 - \kappa_1)(n + 1 - 2\kappa_1) = 0.$$
$$(15.56)$$

It seems likely that a useful approximation could be deduced from this equation, but results are not complete at the time of writing.

A similar approach might be tried in handling the partial differential equations for the competition and prey–predation processes discussed in Chapter 13. Suppose the appropriate equations in (13.9) and (13.22) are transformed into partial differential equations for the cumulant-generating functions. If we equate coefficients of powers of θ_1 and θ_2, neglecting all joint cumulants of order higher than j, say, will provide a closed set of ordinary differential equations for cumulants up to order j. Thus if $j = 2$, we should obtain 5 equations for the 5 unknowns κ_{10}, κ_{01}; κ_{20}, κ_{11}, κ_{02}.

The consequences of such an approximation have so far not been explored.

For a detailed account of the theory of such normal approximations, see Whittle (1957).

15.5 Stochastic linearization

In some situations, when numbers are sufficiently large, it may be possible to investigate stochastic variations about an equilibrium in a manner similar to the treatment of small deterministic oscillations about an equilibrium. Let us consider, for example, Bartlett's (1956) treatment of the recurrent epidemics of Section 12.4. The deterministic version involved the processes of infection and removal, as well as the steady introduction of new susceptibles. And in the stochastic version we also considered the random introduction of new infections from outside to trigger off a new epidemic when the previous one had become extinct. We now suppose that x_0 and y_0 are both large for the equilibrium point (x_0, y_0) given by (12.56) in the deterministic case. Generalizing to the analogous stochastic model, we can ignore the small chance of extinction, and examine the behavior of small stochastic departures from the equilibrium value.

As before, we write $X(t)$ and $Y(t)$ for the random variables representing the numbers of susceptibles and infectives, respectively, at time t. The stochastic change ΔX in $X(t)$ over time t may be regarded as being compounded of a loss due to infection, having a Poisson distribution with parameter $\beta X Y \Delta t$; and a gain due to the influx of new susceptibles, having a Poisson distribution with parameter $\mu \Delta t$. Similarly, the stochastic change ΔY in $Y(t)$ may be regarded as being compounded of a gain due to the infection mentioned above, having the Poisson distribution with parameter $\beta X Y \Delta t$; and a loss due to removal, having a Poisson distribution with parameter $\gamma Y \Delta t$. It is convenient for the subsequent analysis to replace the Poisson variables by new variables with zero means, but with the same variances. We can thus write

$$\left.\begin{aligned} \Delta X &= -\beta X Y \Delta t + \mu \, \Delta t - \Delta Z_1 + \Delta Z_2 \\ \Delta Y &= \beta X Y \Delta t - \gamma Y \Delta t + \Delta Z_1 - \Delta Z_3 \end{aligned}\right\}, \qquad (15.57)$$

where the random variables ΔZ_1, ΔZ_2, and ΔZ_3 are essentially independent Poisson variables, but with means adjusted to zero, having variances $\beta X Y \Delta t$, $\mu \, \Delta t$, and $\gamma Y \Delta t$, respectively. Equations (15.57) are thus approximate stochastic equations corresponding to the strictly deterministic version appearing in (12.55).

We now write

$$X = x_0(1 + U), \qquad Y = y_0(1 + V), \qquad (15.58)$$

for small stochastic departures from the equilibrium position, where $U \equiv U(t)$ and $V \equiv V(t)$ are small random variables. Thus (15.58) is the stochastic equivalent of (12.57). It follows from (15.58) that $\Delta X = x_0 \, \Delta U$ and $\Delta Y = y_0 \, \Delta V$. Thus the *non-linear* stochastic equations in (15.57) can be approximated by retaining only first-order terms in U and V to give the *linear* equations

$$\left. \begin{array}{l} x_0 \, \Delta U = -\mu(U + V)\,\Delta t - \Delta Z_1 + \Delta Z_2 \\[2mm] y_0 \, \Delta V = \mu U \, \Delta t + \Delta Z_1 - \Delta Z_3 \end{array} \right\}, \qquad (15.59)$$

using the fact that $(x_0, y_0) \equiv (\gamma/\beta, \mu/\gamma)$.

The variances of ΔZ_1, ΔZ_2, and ΔZ_3 are now $\mu(1 + U + V)\,\Delta t$, $\mu\,\Delta t$ and $\mu(1 + V)\,\Delta t$, respectively.

Since $\Delta U = U(t + \Delta t) - U(t)$ and $\Delta V = V(t + \Delta t) - V(t)$, we can put (15.59) in the form

$$\left. \begin{array}{l} x_0 U(t + \Delta t) = x_0 U - \mu(U + V)\,\Delta t - \Delta Z_1 + \Delta Z_2 \\[2mm] y_0 V(t + \Delta t) = y_0 V + \mu U \, \Delta t + \Delta Z_1 - \Delta Z_3 \end{array} \right\}. \qquad (15.60)$$

Let us now write $\sigma_U{}^2$, $\sigma_V{}^2$, and σ_{UV} for the variances and covariance of the equilibrium distribution of U and V. As $E(U) = 0 = E(V)$, we have

$$\left. \begin{array}{l} \sigma_U{}^2 = E\{U(t)\}^2 = E\{U(t + \Delta t)\}^2 \\[2mm] \sigma_V{}^2 = E\{V(t)\}^2 = E\{V(t + \Delta t)\}^2 \\[2mm] \sigma_{UV} = E\{U(t)V(t)\} = E\{U(t + \Delta t)V(t + \Delta t)\} \end{array} \right\}. \qquad (15.61)$$

These quantities may now be obtained by squaring and cross-multiplying the equations of (15.60), and then taking expectations. If we retain only terms of first order in Δt, we find that Δt cancels through to yield, after a little reduction,

$$\left. \begin{array}{l} x_0(\sigma_U{}^2 + \sigma_{UV}) = 1 \\[2mm] y_0 \sigma_{UV} = -1 \\[2mm] x_0 \sigma_U{}^2 - y_0(\sigma_{UV} + \sigma_V{}^2) = 1 \end{array} \right\}. \qquad (15.62)$$

We can thus write, for sufficiently small stochastic fluctuations about the equilibrium position,

$$\sigma_U{}^2 = \frac{1}{x_0} + \frac{1}{y_0}, \qquad \sigma_{UV} = -\frac{1}{y_0}, \qquad \sigma_V{}^2 = \frac{1}{y_0} + \frac{x_0}{y_0{}^2}. \qquad (15.63)$$

Similar methods may prove to be useful in other situations in which equilibrium or quasi-equilibrium (i.e. very small chance of extinction) fluctuations are possible.

CHAPTER 16

Some Non-Markovian Processes

16.1 Introduction

Most of the discussions that have been undertaken so far in this book have been subject to the Markov restriction that the future probability behavior of the process in question is uniquely determined once the state of the system at the present stage is given. The restriction is of course mathematically highly convenient. Moreover, a surprisingly large number of real situations can be usefully studied, at least as a first approximation, by means of a suitably chosen Markov chain or process. The restriction is thus not as serious as might at first sight appear.

However, a more accurate representation of reality frequently entails a more general, non-Markovian, formulation. Such models are usually rather difficult to handle. Fortunately, there are a number of devices by means of which it may be possible to restore the Markov property without abandoning the basic model. Thus, as pointed out in Section 11.2, the length of a queue in continuous time, for a queueing process with random arrival but with a general distribution of service-time, does not have the Markov property. Nevertheless, it was shown that the queue lengths left behind by successive departing customers could be used to specify a Markov chain. A number of important results for this Markov chain could then be deduced by relatively simple arguments. It is always worth considering whether it may be possible to derive from a primary non-Markovian process some secondary *imbedded Markov chain*, which will allow of a relatively straightforward analysis by some of the methods already described in this book.

Another technique has already been discussed in Section 10.5 in dealing with a basically non-Markovian birth process involving a χ^2-distribution of generation-time. We used the idea of dividing each life-time into a number of imaginary phases with random transitions from phase to phase. The *multi-dimensional process* for the set of variables representing the

222

numbers of individuals in the several phases was then Markov in character. Standard methods of investigation could be applied, though in this particular case some fairly complicated mathematics developed.

There are other ways in which non-Markovian processes can be converted into Markov processes, such as the *inclusion of supplementary variables*, a method also referred to as the *augmentation technique*. Consider, for example, the single-server queue with random arrivals and general distribution of service-time, already referred to above in connection with the imbedded Markov chain method. Let us examine the joint distribution at any time of the pair of variables specified by (i) the number of customers in the system, and (ii) the length of time that has elapsed since the service of the last customer began. It is easy to see that the two-dimensional process in continuous time for the pair of variables is Markovian in character (see D. R. Cox, 1955, for further discussion). We can therefore proceed on the basis of an analysis which is suitable for Markov processes.

In many situations we may well find that there is no satisfactory reduction of a complicated process to a simpler Markovian form. However, one frequently useful procedure is to use a model which is in effect a generalization of the recurrent events of Chapter 3. It will be remembered that, although we required several of the theorems quoted in Chapter 3 for application to the Markov chains of Chapter 5, the theory of Chapter 3 did not entail the Markov restriction. In essence, we consider some object (which may, for instance, be an industrial product or a living organism) with a general distribution of life-time. When the object's "life" is ended, it is immediately replaced by a new object. We may then inquire about the probability distribution of the total number of replacements that have occurred by any specific time t. The associated stochastic process in continuous time is clearly non-Markovian in general. There is a considerable amount of literature on the general theory of such processes, although the detailed analysis of particular problems may be quite difficult. Section 16.2 below gives an introductory account of the theory of these *recurrent processes* or *renewal processes*, with a specific application to the problem of chromosome mapping in genetics.

An extension of the foregoing ideas may be made analogously to the idea of simplifying the treatment of a non-Markovian process by searching for a simpler imbedded Markov chain. In some cases we may be able to treat an apparently very complicated process by discovering an imbedded renewal process. When this is possible we talk of a *regenerative process*. (There is some variation in the literature about the use of this term, so care should be taken to check definitions.)

In dealing with Markov processes, especially when using generating functions, we have usually had to solve nothing worse than a relatively simple partial differential equation. With non-Markovian processes, however, it is a common occurrence for the basic mathematical description to involve an integral equation. This introduces difficulties of a higher order. Accordingly, Section 16.3 is devoted to a simple introductory discussion of how such equations may be formulated in connection with certain stochastic processes. A full mathematical account is beyond the scope of the present book, and the interested reader should consult the literature cited.

16.2 Renewal theory and chromosome mapping

As mentioned in the previous section, some of the most important types of non-Markovian processes are those known as *recurrent processes* or *renewal processes* (see W. L. Smith, 1958, for an extensive review of modern developments and bibliography). The existing literature is of considerable extent. We shall give here only an elementary introduction to the theory, with special reference to the use of a renewal model to describe some of the phenomena of genetic linkage, especially as an aid to constructing chromosome maps (for a detailed account of this particular application see Chapter 11 of Bailey, 1961).

Let us consider some piece of equipment, e.g. an electric light bulb, which has a certain effective life-time and then needs to be replaced. Let all bulbs have life-times u which are independently distributed with identical frequency distributions $f(u)$, for which the corresponding distribution function is given by

$$F(u) = \int_0^u f(v)\,dv; \qquad (16.1)$$

and let us assume that each bulb is replaced by a new one as soon as it fails. Suppose we now consider a random variable $X(t)$ which represents the number of replacements that have occurred up to time t. Let $p_n(t)$ be the probability that $X(t)$ takes the value n at time t.

Next, let the interval of time elapsing up to the occurrence of the nth renewal have frequency function $f_n(u)$ and distribution function $F_n(u)$. Clearly $f_1(u) \equiv f(u)$. It follows from elementary probability considerations that

$$p_n(t) = F_n(t) - F_{n+1}(t), \quad n \geq 0, \qquad (16.2)$$

where, for convenience, we define $F_0(t) \equiv 1$.

The discussion is very concisely expressed in terms of Laplace transforms. Using the definition already introduced in (11.49), we may write

$$
\left.
\begin{aligned}
f^*(s) &= \int_0^\infty e^{-st} f(t)\, dt \\
f_r^*(s) &= \int_0^\infty e^{-st} f_r(t)\, dt
\end{aligned}
\right\}.
\tag{16.3}
$$

The transform of $F(t)$ is of course given by

$$
F^*(s) = \int_0^\infty e^{-su}\left\{ \int_0^u f(v)\, dv \right\} du = s^{-1} f^*(s),
\tag{16.4}
$$

as follows from integrating by parts. Similarly,

$$
F_n^*(s) = s^{-1} f_n^*(s).
\tag{16.5}
$$

Now the frequency distribution $f_n(u)$ is just the convolution of n independent frequency distributions $f(u)$. It therefore follows that

$$
f_n^*(s) = \{f^*(s)\}^n,
\tag{16.6}
$$

using a standard result in transform theory. Alternatively, we may observe that the Laplace transforms of frequency functions in (16.3) are essentially moment-generating functions. We can then appeal to the well-known result that the moment-generating function of a sum of independent variables is equal to the product of the corresponding individual moment-generating functions.

From (16.5) and (16.6) we have

$$
F_n^*(s) = s^{-1} \{f^*(s)\}^n.
\tag{16.7}
$$

If we now take transforms of both sides of (16.2), we can use (16.7) to write

$$
p_n^*(s) = s^{-1} \{f^*(s)\}^n \{1 - f^*(s)\}, \quad n \geqslant 0.
\tag{16.8}
$$

It is convenient to work with the probability-generating function given by $P(z, t) = \sum_{n=0}^\infty p_n(t) z^n$, with transform $P^*(z, s) = \sum_{n=0}^\infty p_n^*(s) z^n$. Multiplying both sides of (16.8) by z^n, and summing for $0 \leqslant n < \infty$, gives

$$
P^*(z, s) = \frac{1 - f^*(s)}{s\{1 - zf^*(s)\}}.
\tag{16.9}
$$

Transforms of the probabilities, $p_n^*(s)$, are obtained by picking out the appropriate coefficients on the right of (16.9). If we wish to work with the transform of the moment-generating function of the number of replacements in time t, this is immediately obtained from (16.9) as

$$M^*(\theta, s) = \frac{1 - f^*(s)}{s\{1 - e^\theta f^*(s)\}}.$$

(16.10)

In particular, the transform $m^*(s)$ of the mean number of replacements $m(t)$ in time t is easily derived, namely

$$m^*(s) = \frac{f^*(s)}{s\{1 - f^*(s)\}}.$$

(16.11)

We have carried through the above discussion for the simple case in which a frequency density function $f(u)$ exists. More generally, when the frequency density function does not exist, we can replace (16.3) by definitions in terms of Laplace–Stieltjes transforms, viz.

$$\left. \begin{aligned} f^*(s) &= \int_0^\infty e^{-st}\, dF(t) \\ f_r^*(s) &= \int_0^\infty e^{-st}\, dF_r(t) \end{aligned} \right\}.$$

(16.12)

See, for example, Smith (1958), where the notation differs in certain respects from the present one.

It is clear that the renewal process represented by $X(t)$ will in general be non-Markovian, though a number of important properties may not be too difficult to obtain owing to the relative simplicity of such results as (16.9)–(16.11). We shall now examine in a little more detail a specifically biological application.

Chromosome mapping in genetics

In the general theory of chromosome mapping (see Bailey, 1961, Chapter 11, for a more extensive discussion) it is possible to assume that the points of exchange that occur on a single chromosome strand during the appropriate stage of meiosis follow a renewal process of the type discussed above. We suppose that the strand is represented in some suitable scale of measurement, or *metric*, by a semi-infinite straight line, with origin corresponding to the chromosome's centromere. Points of exchange occur in succession along the strand, with the centromere as an obligatory starting point. We suppose that the intervals between successive points of exchange, measured in the units of the metric, are independently distributed with identical frequency functions $f(u)$. The metrical variable t, say, is thus a distance rather than a time. We might equate the two if it were possible to assume that the length of the paired segment increased during the stage of formation of exchange-points at some constant velocity.

However, with the foregoing definitions, we can interpret the renewal theory formulas directly in terms of the phenomena of genetic linkage and chromosome mapping. Thus the *map distance* $x(t)$ corresponding to any metrical length t is simply the average number of points of exchange in the interval $(0, t)$. The transform $x^*(s)$ is accordingly identical with $m^*(s)$ in (16.11). So we can write

$$x^*(s) = \frac{f^*(s)}{s\{1 - f^*(s)\}}. \tag{16.13}$$

Now in practice what we can estimate more or less directly from actual data is the *recombination fraction*, using an appropriately designed linkage backcross experiment. However, as is well known, the recombination fraction is the probability of an *odd* number of points of exchange in the interval $(0, t)$, and is therefore given by

$$y(t) = \sum_{n=0}^{\infty} p_{2n+1}(t) = \tfrac{1}{2}\{1 - P(-1, t)\}, \tag{16.14}$$

for which the corresponding transform is

$$y^*(s) = \tfrac{1}{2}\{s^{-1} - P^*(-1, s)\}$$

$$= \frac{f^*(s)}{s\{1 + f^*(s)\}}, \tag{16.15}$$

using (16.9).

In the special case where points of exchange occur randomly as a Poisson process, the frequency distribution between successive points of exchange will be a negative exponential. Thus, choosing for convenience metrical scale units for which the average distance between two successive points of exchange is unity, we may put $f(u) = e^{-u}$. Thus $f^*(s) = (s + 1)^{-1}$. Substituting in (16.13) then gives $x^*(s) = s^{-2}$, i.e. $x = t$; and substituting in (16.15) gives $y^*(s) = s^{-1}(s + 2)^{-1}$, i.e. $y = \tfrac{1}{2}(1 - e^{-2t})$. Here the special metric is identical with the map distance, and we can write

$$y = \tfrac{1}{2}(1 - e^{-2x}), \tag{16.16}$$

a result usually called Haldane's formula, and easily obtained by more elementary methods.

The Poisson process is of course Markovian, and has already been studied in Section 7.2. Although this process may be used as a first approximation to the occurrence of points of exchange along a chromosome strand, detailed investigation usually shows it to be inadequate. An appreciable amount of non-randomness, or *interference*, can be accounted for by choosing a suitable renewal model. A fair amount of success has been obtained with A. R. G. Owen's metric (see Bailey, 1961, Section 12.2), which essentially involves taking $f(u) = 4ue^{-2u}$. This time $f^*(s) = 4(s + 2)^{-2}$, so that substitution in (16.13) gives $x^*(s) = 4s^{-2}(s + 4)^{-1}$. Inversion yields

$$x(t) = -\tfrac{1}{4} + t + \tfrac{1}{4}e^{-4t}. \tag{16.17}$$

Again, substitution in (16.15) gives $y^*(s) = 4s^{-1}(s^2 + 4s + 8)^{-1}$, from which we can obtain

$$y(t) = \tfrac{1}{2} - \tfrac{1}{2}e^{-2t}(\cos 2t + \sin 2t). \tag{16.18}$$

Here we see how map distance and recombination fraction are connected by means of a parametric representation, in which the variable t refers to the metrical scale used.

In more elaborate studies it is important to consider the more general interval (t_1, t_2) instead of the interval $(0, t)$ starting at the origin, but this is beyond the scope of the present discussion.

16.3   Use of integral equations

Throughout most of this book we have been working with Markov chains or Markov processes, and the generating-function technique has been used extensively. A facility in handling this technique is not difficult to acquire. Equations for the required probability-generating function or moment-generating function are often quite simple in form, and in the more difficult cases may be no worse than some fairly standard type of partial differential equation. On the other hand, in formulating the basic equation for a non-Markovian process it frequently happens that this is most easily done in terms of an integral equation. As a general rule the handling of integral equations involves mathematical difficulties of appreciably greater degree than the treatment of differential and partial differential equations. It is not proposed to deal with the solution of integral equations here, but it is instructive to see how they arise in certain non-Markovian cases that would be difficult to represent in terms of the theory used so far in this book. (For further discussion of these matters the reader should consult Bartlett, 1960; Bharucha-Reid, 1960; etc., *passim*.)

We shall first examine a generalized queueing process, and show how an integral equation for the waiting-time distribution can be derived. Secondly, we shall look at a branching process in which the distribution of individual life-times is of general form. In this case it is fairly easy to obtain an integral equation for the probability-generating function of population size at any time.

Generalized queueing process

Let us consider a single-server queue of the type already discussed in Chapter 11. So far as equilibrium theory was concerned, a number of important results for the case of general distribution of service-time and general distribution of inter-arrival time could be obtained by using the imbedded Markov chain technique. It was shown, in fact, that an appropriate random variable was the length of queue left behind by a departing customer. A more general approach to the non-equilibrium theory was first developed by Lindley (1952). This goes as follows.

We suppose that the queue discipline is "first come, first served"; that the service-time and waiting-time of the rth customer are s_r and w_r; and that the time elapsing between the arrival of the rth and $(r+1)$th customers is t_r. Let us assume the s_r are independent random variables with identical distributions; and that the t_r are also independent random variables with identical distributions. It is further convenient to assume that the s_r and t_r are independently distributed sequences, but this restriction can be somewhat relaxed. If therefore we introduce the variable u_r given by

$$u_r = s_r - t_r, \qquad (16.19)$$

the u_r will be independent random variables with identical distributions. Let the latter have distribution function $G(u)$, and let w_r have distribution function $F_r(w)$.

When the $(r+1)$th customer arrives he may have to wait until the service of the rth customer is completed. In this case we have $t_r < w_r + s_r$ and $t_r + w_{r+1} = w_r + s_r$. If, on the other hand, the $(r+1)$th customer does not have to wait, we have $t_r \geqslant w_r + s_r$ and $w_{r+1} = 0$. These relations can be written as

$$\left. \begin{aligned} w_{r+1} &= w_r + u_r, \qquad w_r + u_r > 0 \\ &= 0, \qquad\qquad w_r + u_r \leqslant 0 \end{aligned} \right\}. \qquad (16.20)$$

Next, we can argue from simple probability considerations that

$$\begin{aligned} F_{r+1}(w) &= P\{w_{r+1} \leqslant w\} \\ &= P\{w_{r+1} = 0\} + P\{0 < w_{r+1} \leqslant w\} \\ &= P\{w_r + u_r \leqslant 0\} + P\{0 < w_r + u_r \leqslant w\} \\ &= P\{w_r + u_r \leqslant w\} \\ &\equiv \int_{w_r + u_r \leqslant w} dF_r(w_r)\, dG(u_r) \\ &= \int_{u_r \leqslant w} F_r(w - u_r)\, dG(u_r). \qquad (16.21) \end{aligned}$$

Now if the first customer does not have to wait, we have $F_1(w) = 1$, $w \geqslant 0$. Thus for $r \geqslant 1$, equation (16.21) enables us to determine the $F_r(w)$ successively. It is interesting to observe that the form of (16.21) implies that the waiting-time distribution of any customer depends only on the distribution of the u_r, i.e. on the distribution of the *difference* between service-time and inter-arrival time, and not on their individual distributions.

It can be shown that, for an equilibrium distribution of waiting-time to exist as $r \to \infty$, we must in general have $E(u) < 0$ (as we should expect from the discussion of Section 11.2). From (16.20) it follows that the equilibrium distribution $F(w)$ must satisfy

$$F(w) = \int_{u \leqslant w} F(w - u) \, dG(u)$$

$$= \int_{v \geqslant 0} F(v) \, dG(w - v). \qquad (16.22)$$

If a frequency function $g(u)$ exists for the variables u_r, equation (16.22) can be written in the form

$$F(w) = \int_0^\infty F(v) g(w - v) \, dv. \qquad (16.23)$$

This integral equation, which is of Wiener–Hopf type, is somewhat difficult to handle in general, though a number of special cases have been successfully investigated (see, for example, Saaty, 1961, Section 9.3).

Age-dependent branching processes

In the discussion of a simple birth process in Section 8.2 we made the simplifying assumption that the chance of a given individual producing a new one (or, alternatively, being *replaced* by two new individuals) in time Δt was $\lambda \, \Delta t$. With a single-celled organism, like a bacterium, this was equivalent to assuming a negative exponential distribution of the time elapsing between the birth of an individual and its subsequent division into two daughter cells. These assumptions were highly convenient, as they enabled us to treat the development of a whole population as a Markov process in continuous time. With a more general distribution of individual life-times a basic equation describing the process is less easily formulated than the simple partial differential equation previously derived in (8.4).

Let us consider a branching process in continuous time for which the individual life-times are all independent with identical distribution functions $F(u)$. For a Markov process we should have $F(u) = 1 - e^{-\lambda u}$, so that there would then be a negative exponential frequency function

$f(u) = \lambda e^{-\lambda u}$, as in (8.1). We can also make a further generalization in the present case, and assume that at the end of an individual life-time the organism is replaced by j daughter cells with probability $q_j, j = 0, 1, 2, \cdots$. Let the associated probability-generating function be $Q(x) = \sum_{j=0}^{\infty} q_j x^j$. In the Markov case with division into just two daughter cells we should simply have $q_2 = 1; q_j = 0, j \neq 2$.

As usual we represent the total number of individuals in the population at time t by the random variable $X(t)$, where the probability that $X(t)$ takes the value n is $p_n(t)$, and the corresponding probability-generating function is $P(x, t) = \sum_{n=0}^{\infty} p_n(t) x^n$.

Let us examine a population starting with a single individual whose life-time ends at time τ with probability $dF(\tau)$. Now, given that this individual is replaced at time τ, the probability-generating function of the population size at $t \geqslant \tau$ is

$$P(x, t) = Q(P(x, t - \tau)), \qquad (16.24)$$

applying formula (2.27) to the present example of a compound distribution.

Alternatively, the chance that the individual has not been replaced by time t is $1 - F(t)$. In this case, when $\tau > t$, the probability-generating function of total population size is simply x. It is clearly legitimate to collect together all the contributions for $\tau \leqslant t$ and $\tau > t$, to give

$$P(x, t) = \int_0^t Q(P(x, t - \tau)) \, dF(\tau) + x\{1 - F(t)\}. \qquad (16.25)$$

If any doubt is felt about the validity of the above argument, a more extended proof can easily be developed along the lines of Section 2.4 in which the various probability statements are set out explicitly and then condensed into an equation involving generating functions.

Equation (16.25) is a very useful starting point for further investigation, even if it proves somewhat intractable in particular cases. We can, for instance, obtain an integral equation for the mean population size $m(t)$ at time t by differentiating (16.25) with respect to x, and then putting $x = 1$. Thus

$$m(t) = \frac{\partial P(x, t)}{\partial x}\bigg|_{x=1}$$

$$= \mu \int_0^t m(t - \tau) \, dF(\tau) + 1 - F(t), \qquad (16.26)$$

where $\mu = [dQ/dy]_{y=1}$ is the mean number of daughter cells.

In a similar way we can obtain an integral equation for the variance $\sigma^2(t)$ of the population size at time t by differentiating (16.25) twice with respect to x, and then putting $x = 1$.

If the distribution of life-times has a frequency function $f(u)$, we can write (16.25) in the alternative form

$$P(x, t) = \int_0^t Q\{P(x, t - \tau)\}f(\tau)\,d\tau + x\{1 - F(t)\}. \qquad (16.27)$$

It may be instructive to examine briefly one or two special cases. First, consider the simple birth process of Section 8.2, for which $F(u) = 1 - e^{-\lambda u}$ and $f(u) = \lambda e^{-\lambda u}$; and $q_2 = 1$, with $q_j = 0$ for $j \neq 2$. Equation (16.27) then becomes

$$P(x, t) = \int_0^t \{P(x, t - \tau)\}^2 \lambda e^{-\lambda \tau}\,d\tau + xe^{-\lambda t}. \qquad (16.28)$$

If we now change the variable of integration from τ to v, where $t - \tau = v$, and multiply the whole equation by $e^{\lambda t}$, we obtain

$$e^{\lambda t} P(x, t) = \int_0^t \{P(x, v)\}^2 \lambda e^{\lambda v}\,dv + x.$$

Differentiation with respect to t now yields, after rearrangement,

$$\frac{\partial P}{\partial t} = \lambda P(P - 1). \qquad (16.29)$$

This equation is essentially what is obtained by the procedure already described in Section 8.9 for a multiplicative Markov process. Thus (16.29) is identical with the backward equation (8.95), with $\mu = 0$.

Suppose we now consider, more generally, the simple birth-and-death process of Section 8.6. We have $F(u) = 1 - e^{-(\lambda + \mu)u}$, since the distribution of the time from birth to the occurrence of the first transition (i.e. giving birth or dying) is negative exponential with parameter $\lambda + \mu$. Next, the q_j probabilities are obviously given by

$$q_0 = \frac{\mu}{\lambda + \mu}, \quad q_1 = 0, \quad q_2 = \frac{\lambda}{\lambda + \mu};$$

since, conditionally, a particle either dies or is replaced by two daughter cells. Equation (16.27) then takes the form

$$P(x, t) = \int_0^t [\mu + \lambda\{P(x, t - \tau)\}^2]e^{-(\lambda + \mu)\tau}\,d\tau + xe^{-(\lambda + \mu)t}. \qquad (16.30)$$

Proceeding as before, we obtain the partial differential equation

$$\frac{\partial P}{\partial t} = \lambda P^2 - (\lambda + \mu)P + \mu, \qquad (16.31)$$

which is identical with equation (8.95).

Of course, in the above illustrations we have used very simple processes which are in fact Markovian and are accordingly treated quite easily by other means. In non-Markov cases the methods of the present discussion provide a very powerful tool for investigating processes that might otherwise seem intractable.

References

The following list of references is intended primarily as a guide to further reading, and not as an exhaustive bibliography of the vast literature on stochastic processes. Many of the books included below do contain extensive bibliographies, and these have been specially indicated in the text. A number of books on mathematical theory have also been added, where these treat special topics used in the handling of stochastic processes.

ARLEY, N. (1943). *On the Theory of Stochastic Processes and Their Application to the Theory of Cosmic Radiation.* Copenhagen.

ARMITAGE, P. (1951). The statistical theory of bacterial populations subject to mutation, *J. R. Statist. Soc.*, **B 14,** 1.

BAILEY, N. T. J. (1954). Queueing for medical care, *Applied Statistics*, **3,** 137.

—— (1957). *The Mathematical Theory of Epidemics.* London: Griffin; New York: Hafner.

—— (1961). *The Mathematical Theory of Genetic Linkage.* Oxford: Clarendon Press.

—— (1963). The simple stochastic epidemic: a complete solution in terms of known functions, *Biometrika*, **50,** 11.

BARTLETT, M. S. (1955). *An Introduction to Stochastic Processes.* Cambridge University Press.

—— (1956). Deterministic and stochastic models for recurrent epidemics, *Proc. Third Berkeley Symp. on Math. Stat. and Prob.*, **4,** 81.

—— (1957). On theoretical models for competitive and predatory biological systems, *Biometrika*, **44,** 27.

—— (1960). *Stochastic Population Models in Ecology and Epidemiology.* New York: John Wiley; London: Methuen.

BHARUCHA-REID, A. T. (1960). *Elements of the Theory of Markov Processes and their Applications.* New York: McGraw-Hill.

BRITISH ASSOCIATION MATHEMATICAL TABLES (1950, 1952). Vols. VI and X: *Bessel Functions.* Cambridge University Press.

BUSH, R. R., and MOSTELLER, F. (1951). A mathematical model for simple learning, *Psychol. Rev.*, **58,** 313.

CHANDRASEKHAR, S. (1943). Stochastic problems in physics and astronomy, *Rev. Mod. Phys.*, **15,** 1.

CLARKE, A. B. (1956). A waiting line process of Markov type, *Ann. Math. Statist.*, **27,** 452.

COPSON, E. T. (1948). *An Introduction to the Theory of Functions of a Complex Variable.* London: Oxford University Press.

Cox, D. R. (1955). The analysis of non-Markovian stochastic processes by the inclusion of supplementary variables, *Proc. Cambridge Phil. Soc.*, **51**, 433.

—— (1962). *Renewal Theory*. London: Methuen; New York: Wiley.

—— and Smith, W. L. (1961). *Queues*. London: Methuen; New York: Wiley.

Daniels, H. E. (1954). Saddle-point approximations in statistics, *Ann. Math. Statist.*, **25**, 631.

—— (1960). Approximate solutions of Green's type for univariate stochastic processes, *J. R. Statist. Soc.*, **B 22**, 376.

DeBruijn, N. G. (1958). *Asymptotic Methods in Analysis*. Amsterdam: North Holland Publishing Co.

Doob, J. L. (1953). *Stochastic Processes*. New York: Wiley.

Erdélyi, A. (Ed.) (1954). *Tables of Integral Transforms*, Vol. 1. New York: McGraw-Hill.

Feller, W. (1957). *An Introduction to Probability Theory and its Applications*, Vol. 1 (2nd Ed.). New York: Wiley.

Fisher, R. A. (1949). *The Theory of Inbreeding*. Edinburgh: Oliver & Boyd.

Forsyth, A. R. (1929). *A Treatise on Differential Equations* (6th Ed.). London: Macmillan.

Frazer, R. A., Duncan, W. J., and Collar, A. R. (1946). *Elementary Matrices*. Cambridge University Press.

Furry, W. H. (1937). On fluctuation phenomena in the passage of high energy electrons through lead, *Phys. Rev.*, **52**, 569.

Gause, G. F. (1934). *The Struggle for Existence*. Baltimore: Williams & Wilkins.

Goldberg, S. (1958). *Introduction to Difference Equations*. New York: Wiley.

Haskey, H. W. (1954). A general expression for the mean in a simple stochastic epidemic, *Biometrika*, **41**, 272.

Kelly, C. D., and Rahn, O. (1932). The growth rate of individual bacterial cells, *J. Bacteriol.*, **23**, 147.

Kemeny, J. G., and Snell, L. (1960). *Finite Markov Chains*. New York: Van Nostrand.

Kendall, D. G. (1948a). On the role of variable generation time in the development of a stochastic birth process, *Biometrika*, **35**, 316.

Kendall, D. G. (1948b). On the generalized "birth-and-death" process, *Ann. Math. Statist.*, **19**, 1.

Lea, D. E., and Coulson, C. A. (1949). The distribution of the number of mutants in bacterial populations, *J. Genet.*, **49**, 264.

Lindley, D. V. (1952). The theory of queues with a single server, *Proc. Cambridge Phil Soc.*, **48**, 277.

Lotka, A. J. (1925). *Elements of Physical Biology*. Baltimore: Williams & Wilkins.

McLachlan, N. W. (1953). *Complex Variable Theory and Transform Calculus*. Cambridge University Press.

Molina, E. C. (1942). *Poisson's Exponential Binomial Limit*. New York: Van Nostrand.

Moran, P. A. P. (1962). *The Statistical Processes of Evolutionary Theory*. Oxford: Clarendon Press.

Park, T. (1954). Experimental studies of interspecies competition. II. Temperature, humidity and competition in two species of *Tribolium*, *Physiol. Zool.*, **27**, 177.

—— (1957). Experimental studies of interspecies competition. III. Relation of initial species proportion to competitive outcome in populations of *Tribolium*, *Physiol. Zool.*, **30**, 22.

PARZEN, E. (1960). *Modern Probability Theory and its Applications.* New York: Wiley.
—— (1962). *Stochastic Processes.* San Francisco: Holden-Day.
ROSENBLATT, M. (1962). *Random Processes.* Oxford University Press.
ROTHSCHILD, Lord (1953). A new method of measuring the activity of spermatozoa, *J. Exptl. Biol.*, **30**, 178.
SAATY, T. L. (1961). *Elements of Queueing Theory.* New York: McGraw-Hill.
SAGAN, H. (1961). *Boundary and Eigenvalue Problems in Mathematical Physics.* New York: Wiley.
SMITH, W. L. (1958). Renewal theory and its ramifications, *J. R. Statist. Soc.*, **B 20**, 243.
TAKÁCS, L. (1960). *Stochastic Processes.* New York: Wiley; London: Methuen.
VOLTERRA, V. (1931). *Leçons sur la théorie mathématique de la lutte pour la vie.* Paris: Gauthier-Villars.
WHITTLE, P. (1952). Certain nonlinear models of population and epidemic theory, *Skand. Aktuar.*, **14**, 211.
—— (1955). The outcome of a stochastic epidemic—a note on Bailey's paper, *Biometrika*, **42**, 116.
—— (1957). On the use of the normal approximation in the treatment of stochastic processes, *J. R. Statist. Soc.*, **B 19**, 268.
WILLIAMS, G. T. (1963). The simple epidemic curve for large populations of susceptibles [Unpublished].
YULE, U. (1924). A mathematical theory of evolution based on the conclusions of Dr. J. C. Willis, F.R.S., *Phil. Trans.*, **B 213**, 21.

Solutions to the Problems

Chapter 2

2. $e^{\lambda(x-1)}$.

3. $p/(1-qx)$; q/p; q/p^2.

5. $0, npq, npq(q-p), 3n^2p^2q^2 + pqn(1-6pq)$.

6. $p^n/(1-qx)^n$; nq/p; nq/p^2.

Chapter 3

1. $U(x) = 1 + px/(1-x) = (1-qx)/(1-x)$;
   $F(x) = px/(1-qx)$.

2. $u_n = 2^{-2n} \sum\limits_{r=0}^{n} \binom{n}{r}^2 \sim (n\pi)^{-\frac{1}{2}}$;

   hence Σu_n diverges, but $u_n \to 0$, as $n \to \infty$.

3. $t = 6$, $u_{6n} = 6^{-6n}(6n!)/(n!)^6 \sim Cn^{-5/2}$;
   hence Σu_{6n} converges, and E is transient.

4. $S_n = 0$ can occur only if $n = k(a+b)$, when $u_n \sim (a+b)^{\frac{1}{2}}(2\pi abk)^{-\frac{1}{2}}$.
   Thus Σu_n diverges and the event is persistent.

5. $U(x) = \frac{1}{2} + \frac{1}{2}(1-x^2)^{-\frac{1}{2}}$;
   $F(x) = x^{-2}\{1 - (1-x^2)^{\frac{1}{2}}\}^2$.

Chapter 4

2. $q_k = 1$ if $q \geqslant 2p$;

$$= \left\{ \left(\frac{1}{4} + \frac{q}{p} \right)^{\frac{1}{2}} - \frac{1}{2} \right\}^k \text{ if } q \leqslant 2p.$$

3. $q_k = \dfrac{(1-\delta)\{(q/p)^a - (q/p)^k\}}{(1-\delta)(q/p)^a + \delta(q/p) - 1}$.

4. $q_{k,n+1}(\xi) = pq_{k+1,n}(\xi) + qq_{k-1,n}(\xi)$,
   subject to $q_{0n}(\xi) = q_{an}(\xi) = 0$ if $n \geqslant 1$;
   $q_{k0}(\xi) = 0$ for $k \neq \xi$; $q_{\xi 0}(\xi) = 1$.

5. Consider $E(r_{n+1}^2 - r_n^2)$.

Chapter 5

1.
$$\begin{bmatrix} 1 & q & 0 & \cdots & 0 & 0 \\ 0 & 0 & q & \cdots & 0 & 0 \\ 0 & p & 0 & \cdots & 0 & 0 \\ 0 & 0 & p & \cdots & 0 & 0 \\ \vdots & \vdots & \vdots & & \vdots & \vdots \\ 0 & 0 & 0 & \cdots & q & 0 \\ 0 & 0 & 0 & \cdots & 0 & 0 \\ 0 & 0 & 0 & \cdots & p & 1 \end{bmatrix}$$

2.
$$\begin{bmatrix} q & q & q & \cdots \\ p & 0 & 0 & \cdots \\ 0 & p & 0 & \cdots \\ 0 & 0 & p & \cdots \\ 0 & 0 & 0 & \cdots \\ \vdots & \vdots & \vdots & \end{bmatrix}$$

3. (i) Irreducible and ergodic.
   (ii) Chain has period 3.
   (iii) Two aperiodic closed sets (E_1, E_2) and (E_3, E_4), with E_5 transient.

4. Derive generating functions from equation (5.16).

5.
$$\begin{bmatrix} q & q & 0 & \cdots & 0 & 0 \\ p & 0 & q & \cdots & 0 & 0 \\ 0 & p & 0 & \cdots & 0 & 0 \\ 0 & 0 & p & \cdots & 0 & 0 \\ \vdots & \vdots & \vdots & & \vdots & \vdots \\ 0 & 0 & 0 & \cdots & q & 0 \\ 0 & 0 & 0 & \cdots & 0 & q \\ 0 & 0 & 0 & \cdots & p & p \end{bmatrix} ; \quad p_j = \frac{\{1 - (p/q)\}(p/q)^{j-1}}{1 - (p/q)^{a-1}}, \quad j = 1, 2, \cdots, a-1.$$

6. $\mathbf{P}^n = \begin{bmatrix} \frac{1}{2} & \frac{1}{2} \\ \frac{1}{2} & \frac{1}{2} \end{bmatrix} + (p-q)^n \begin{bmatrix} \frac{1}{2} & -\frac{1}{2} \\ -\frac{1}{2} & \frac{1}{2} \end{bmatrix},$

$\left. \begin{array}{l} p_1^{(n)} = \frac{1}{2} + \frac{1}{2}(\alpha - \beta)(p-q)^n \\ p_2^{(n)} = \frac{1}{2} - \frac{1}{2}(\alpha - \beta)(p-q)^n \end{array} \right\},$

$p_1 = p_2 = \frac{1}{2}.$

Chapter 6

3. $\operatorname{cov}(X_j, X_k) = \dfrac{m^{k-1}(m^j - 1)}{m - 1} \sigma^2.$

Chapter 7

3. $E\{X(t)\} = \lambda t,\ \operatorname{var}\{X(t)\} = \lambda t$

$\begin{aligned} E\{X(t)X(t+\tau)\} &= E[\{X(t)\}^2] + E[X(t)\{X(t+\tau) - X(t)\}] \\ &= E[\{X(t)\}^2] + E\{X(t)\}E\{X(t+\tau) - X(t)\} \\ &= \lambda t + (\lambda t)^2 + (\lambda t)(\lambda \tau) \\ &= \lambda t(\lambda t + \lambda \tau + 1), \end{aligned}$

noting that $X(t)$ and $X(t+\tau) - X(t)$ are independent.

Hence required correlation coefficient is $\left(\dfrac{t}{t+\tau} \right)^{\frac{1}{2}}.$

5. $\dfrac{dp_n}{dt} = \lambda(n-1)p_{n-1} - \lambda n p_n, \quad p_a(0) = 1$.

$$\frac{\partial P}{\partial t} = \lambda x(x-1)\frac{\partial P}{\partial x}, \quad P(x, 0) = x^a.$$

$$p_n(t) = \binom{n-1}{a-1}e^{-a\lambda t}(1-e^{-\lambda t})^{n-a}, \quad n \geq a.$$

Chapter 8

1. $\dfrac{\partial M}{\partial t} = \mu(e^{-\theta}-1)\dfrac{\partial M}{\partial \theta} + \nu(e^{\theta}-1)M, \, M(\theta, 0) = e^{a\theta}$.

Solving gives

$$M(\theta, t) = \{1 - (1-e^{\theta})e^{-\mu t}\}^a \, \exp\left\{-\frac{\nu}{\mu}(1-e^{\theta})(1-e^{-\mu t})\right\};$$

$$P(x, t) = \{1 - (1-x)e^{-\mu t}\}^a \, \exp\left\{-\frac{\nu}{\mu}(1-x)(1-e^{-\mu t})\right\}.$$

2. $K(\theta, t) = \log M(\theta, t)$;

$$m(t) = \left[\frac{\partial K}{\partial \theta}\right]_{\theta=0} = \frac{\nu}{\mu} + \left(a - \frac{\nu}{\mu}\right)e^{-\mu t}.$$

3. $\dfrac{\partial M}{\partial t} = (e^{-\theta}-1)(\mu + \lambda e^{\theta})\dfrac{\partial M}{\partial \theta} + N\lambda(e^{\theta}-1)M$.

4. Put $\dfrac{\partial M}{\partial t} = 0$ and solve $(\mu + \lambda e^{\theta})\dfrac{\partial M}{\partial \theta} = N\lambda e^{\theta}M$,

subject to $M(0) = 1$. We obtain the binomial distribution (as expected) given by

$$M(\theta) = \left(\frac{\mu + \lambda e^{\theta}}{\mu + \lambda}\right)^N.$$

5. Put $K = \log M$. Equation for K is

$$\frac{\partial K}{\partial t} = (e^{-\theta}-1)(\mu + \lambda e^{\theta})\frac{\partial K}{\partial \theta} + N\lambda(e^{\theta}-1).$$

Equate coefficients of θ to obtain

$$\frac{d\kappa_1}{dt} = -(\mu + \lambda)\kappa_1 + N\lambda,$$

where $\kappa_1(0) = a.$ Solving gives

$$m(t) \equiv \kappa_1(t) = \frac{N\lambda}{\mu + \lambda} + \left(a - \frac{N\lambda}{\mu + \lambda}\right) e^{-(\mu+\lambda)t}.$$

7. (i) $\dfrac{dp_{ij}(t)}{dt} = -\lambda_j p_{ij}(t) + \lambda_{j-1} p_{i,j-1}(t);$

$\dfrac{dp_{ij}(t)}{dt} = -\lambda_i p_{ij}(t) + \lambda_i p_{i+1,j}(t).$

(ii) $\dfrac{dp_{ij}(t)}{dt} = \lambda_{j-1} p_{i,j-1}(t) - (\lambda_j + \mu_j) p_{ij}(t) + \mu_{j+1} p_{i,j+1}(t);$

$\dfrac{dp_{ij}(t)}{dt} = \lambda_i p_{i+1,j}(t) - (\lambda_i + \mu_i) p_{ij}(t) + \mu_i p_{i-1,j}(t).$

Chapter 9

2. $\dfrac{\partial K}{\partial t} = \{\lambda(t)(e^\theta - 1) + \mu(t)(e^{-\theta} - 1)\} \dfrac{\partial K}{\partial \theta} + \nu(t)(e^\theta - 1).$

Hence solve

$\kappa'_1 = (\lambda - \mu)\kappa_1 + \nu,$

$\kappa'_2 = 2(\lambda - \mu)\kappa_2 + (\lambda + \mu)\kappa_1 + \nu,$

where λ, μ and ν are all functions of t.

Chapter 10

2. If X represents the normal population, and Y the mutant population, we have

$f_{10} = \lambda(1 - p)X,\ \ f_{01} = \lambda Y + \lambda p X,\ \ $ etc.

Chapter 11

1. $\dfrac{dp_n}{dt} = \lambda_{n-1} p_{n-1} - (\lambda_n + \mu_n) p_n + \mu_{n+1} p_{n+1},\ \ n \geqslant 1;$

$\dfrac{dp_0}{dt} = -\lambda_0 p_0 + \mu_1 p_1.$

2. $p_n = \dfrac{\lambda_0 \lambda_1 \cdots \lambda_{n-1}}{\mu_1 \mu_2 \cdots \mu_n} p_0,\ \ n = 1, 2, \cdots;$

$1 + \dfrac{\lambda_0}{\mu_1} + \dfrac{\lambda_0 \lambda_1}{\mu_1 \mu_2} + \dfrac{\lambda_0 \lambda_1 \lambda_2}{\mu_1 \mu_2 \mu_3} + \cdots$ must converge.

4. $p_n = \dfrac{\rho^n(1-\rho)}{1-\rho^{N+1}}, \quad n = 0, 1, \cdots, N; p_N.$

5. $p_n = \dfrac{\rho^n}{n!} \bigg/ \displaystyle\sum_{n=0}^{N} \dfrac{\rho^n}{n!}, \quad n = 0, 1, \cdots, N.$

Chapter 12

1. The distribution function of T is $p_0(T)$, and the frequency function is dp_0/dT. Thus

$$M_T(\theta) = E(e^{\theta T}) = \int_0^\infty e^{\theta T} \frac{dp_0}{dT}\, dT$$

$$= [e^{\theta T} p_0]_0^\infty - \theta \int_0^\infty e^{\theta T} p_0\, dT, \quad \text{integrating by parts,}$$

$$= -\theta q_0(-\theta), \qquad\qquad \text{from equation (12.11).}$$

Now substitute from (12.13), etc.

2. We can develop a Taylor expansion of μ'_1 about $\tau = 0$, the appropriate coefficients being derived from the set of equations in (12.16) by a process of repeated differentiation and substitution. Unfortunately, no expression is available for the general term.

3. $P_0 = \dfrac{\rho}{\rho+3}, \qquad P_1 = \dfrac{3\rho^2}{(\rho+3)(\rho+2)^2}, \qquad P_2 = \dfrac{6\rho^3(2\rho+3)}{(\rho+3)(\rho+2)^2(\rho+1)^3}$

$P_3 = \dfrac{6(5\rho^3 + 12\rho^2 + 8\rho + 2)}{(\rho+3)(\rho+2)^2(\rho+1)^3}.$

4.

	(a)	Reed–Frost; Greenwood.	(b)	Reed–Frost;	Greenwood
	(1)	q^2	(1)	q^3	q^3
	(1, 1)	$2pq^2$	(1, 1)	$3pq^4$	$3pq^4$
	(1, 1, 1)	$2p^2q$	(1, 1, 1)	$6p^2q^4$	$6p^2q^4$
	(1, 2)	p^2	(1, 2)	$3p^2q^3$	$3p^2q^2$
			(1, 1, 1, 1)	$6p^3q^3$	$6p^3q^3$
			(1, 1, 2)	$3p^3q^2$	$3p^3q^2$
			(1, 2, 1)	$3p^3q(1+q)$	$3p^3q$
			(1, 3)	p^3	p^3

Author Index

Arley, N., 114
Armitage, P., 126

Bailey, N. T. J., 2, 137, 147, 156, 162, 166–168, 175, 184, 224, 226, 228
Bartlett, M. S., 4, 88, 126, 162, 186, 188, 192, 193, 219, 228
Bharucha-Reid, A. T., 4, 101, 107, 119, 137, 201, 204, 228
Bush, R. R., 51

Chandrasekhar, S., 100
Clarke, A. B., 161
Collar, A. R., 3, 50
Copson, E. T., 3, 12, 33, 214
Coulson, C. A., 129
Cox, D. R., 21, 137, 151, 223

Daniels, H. E., 207, 214
Darwin, C., 1
DeBruijn, N. G., 4, 214
Doob, J. L., 3
Duncan, W. J., 3, 50

Erdélyi, A., 153

Feller, W., 3, 5, 12, 16, 18, 19, 21, 22, 27, 34, 38, 42, 51, 90
Fisher, R. A., 56
Forsyth, A. R., 4, 74, 209
Foster, F. G., 47
Frazer, R. A., 3, 50
Frost, W. H., 183
Furry, W. H., 87

Gause, G. F., 192, 193
Goldberg, S., 26
Greenwood, M., 183

Haskey, H. W., 168
Heraclitus, 1

Kelly, C. D., 131
Kemeny, J. G., 38
Kendall, D. G., 121, 125, 131–134
Kermack, W. O., 172, 176

Lea, D. E., 129
Lindley, D. V., 229
Lotka, A. J., 186

McKendrick, A. G., 172, 176
McLachlan, N. W., 3
Molina, E. C., 146
Moran, P. A. P., 201
Mosteller, F., 51

Newton, Isaac, 1

Park, T., 188
Parzen, E., 3, 4, 13
Pólya, G., 34

Rahn, O., 131
Reed, L. J., 183
Rosenblatt, M., 4
Rothschild, Lord, 100

Saaty, T. L., 137, 230
Sagan, H., 203
Smith, W. L., 137, 151, 224, 226
Snell, L., 38
Soper, H. E., 179

Takács, L., 4

Volterra, V., 186

Whittle, P., 177, 219
Williams, G. T., 168

Yule, U., 87

Subject Index

Absolute probabilities, 39
Absorbing barrier, 24, 176
Accidents, 68
Additive process, 202
Age-dependent branching process, 230
Aperiodic Markov chain, 47
Aperiodic state, 44
Appointment systems, in hospital out-
    patient departments, 147
Approximations 207 ff.
    asymptotic, in queueing theory, 155,
        156
    continuous, 207
    neglect of high-order cumulants, 217
    saddle-point, 214
    steepest descents, 214
    stochastic linearization, 219
Asymptotic approximations, 4
    in queueing theory, 155, 156
Augmentation technique, 223

Backward equation
    for diffusion process, 200, 201
    for Markov process, 78
Bacteria, mutation in, 125
Bacterium aerogenes, 131
Barriers in random walk
    absorbing, 24, 176
    elastic, 25
    reflecting, 25
Birth-and-death process, 3, 84 ff.
    effect of immigration in, 97, 115
    general, 101
    homogeneous, 84 ff.
    non-homogeneous, 110
    simple, 91, 232
Birth process
    diffusion approximation for, 210
    divergent, 89
    explosive, 89
    general, 88

multiple-phase, 131
simple, 84, 232
Branching process, 10
    age-dependent, 230
    continuous approximation to, 207
    discrete, 58 ff.
Brownian motion, of colloidal par-
    ticles, 100, 194

Chain-binomial models, in epidemic
    theory, 182
Chance of adequate contact, in epi-
    demics, 183
Chapman–Kolmogorov equations, 76
    for diffusion processes, 199
    for Markov chains, 76
Characteristic function, 13
Chromosomes
    breakage of, 66, 68
    mapping of, 226
Colloidal particles, motion of, 100, 194
Competition
    and predation, 186 ff.
    between two species, 187
Compound distribution, 9
Conditional probabilities, 39
Continuous variables in continuous
    time, 194
Convolution, 7
Cumulant-generating function, 7, 12
Cumulants
    generating function for, 7, 12
    neglect of high order, 217
Cumulative population, treatment of,
    120

Death process
    general, 91
    simple, 90
Delayed recurrent events, 21, 45
Deterministic growth, 87, 89, 94

Deterministic model, 2
   in competition between species,
      187
   in epidemics, 164, 170, 177
   in prey–predator relationship, 189
Deterministic threshold theorem, in
      epidemics, 172
Differential-difference equations, 67
Differential equations, 3
   for stochastic moments, 96
   Kolmogorov backward, 78
   Kolmogorov forward, 77
   solution of partial, 74, 202
Diffusion coefficient, 196
Diffusion equations, 197, 199–201
   singular, 201
   solution of, 202
Diffusion processes, 194 ff.
   additive, 202
   application to population growth,
      205
   as limit of discrete branching pro-
      cesses, 197
   as limit of random walk, 195
   general theory of, 199
   homogeneous, 202
Discrete variables
   in continuous time, 66 ff.
   in discrete time, 16–65
Divergent birth process, 89
Divergent growth, 80
Drift, in diffusion process,, 196
Duration of game, in gambler's ruin
      problem, 27
Duration time, of simple stochastic
      epidemic, 168

Eigenfunction, 203
Eigenvalue, 47, 203
Eigenvector, 47
Elastic barrier, 25
Epidemic processes, 2, 162 ff.
   chain-binomial models for, 182
   deterministic, 164, 170, 177
   general, 169
   recurrent, 177
   simple, 164
   stochastic, 165, 172, 179
Ergodic process, 80
Ergodic state, 44

Event, 16
   delayed recurrent, 21
   periodic, 18
   persistent, 18
   recurrent, 16
   transient, 18
Evolution, 1, 87
Evolutionary process, 4
Explosive birth process, 89
Explosive growth, 61, 80
Exponential distribution (see Negative
      exponential distribution)
Extinction, probability of, 60, 95

Factorial moment-generating function,
      7, 14
Factorial moments, generating func-
      tion for, 7, 14
Family names, survival of, 59
First-passage time, 30
   in diffusion processes, 204
   in queueing theory, 157
Fokker–Planck equation, 197, 201
Forward equation
   for diffusion process, 199–201
   for Markov process, 77
Fourier transform, 13

Gambler's ruin, 25
   expected duration of game, 27
   probability of ruin at nth trial, 28
   probability of ruin over-all, 25
General birth-and-death process, 101
General birth process, 88
General death process, 91
General epidemics, 169
Generating function, 3, 5 ff.
   cumulant, 7, 12
   factorial moment, 7, 14
   moment, 7, 12
   probability, 6
Genetic linkage, 227
Greenwood's chain binomial, 183

Homogeneous birth-and-death pro-
      cess, 84 ff.
Homogeneous Markov chain, 40
Hospital beds, demand for, 143
Hospitals, function and design of, 136

Imbedded Markov chain, 138, 222
Immigration, effect of, on birth-and-death process, 97, 115
Inbreeding, 53
Incubation period, 163
Infectious period, in epidemics, 163
Input process, of a queue, 137
Integral equations, use of, 228
Interaction between species, 186
Interference, in genetics, 227
Irreducible Markov chain, 42

Kermack and McKendrick's Threshold Theorem, 172
Kolmogorov
    backward differential equation, 78
    backward diffusion equation, 200, 201
    forward differential equation, 77
    forward diffusion equation, 199, 201

Lagrange's formula, for the reversion of series, 33
Laplace transform 3, 142
    for solving differential-difference equations, 151, 166
    for solving partial differential equations, 203
Latent period, in epidemics, 163
Linearization, stochastic, 219

Map distance, in genetics, 227
Markov chain, 38 ff.
    aperiodic, 47
    classification of chains, 46
    classification of states, 41
    ergodic, 46
    finite, 40, 46
    homogeneous, 40
    imbedded, 138
    irreducible, 42
    limiting distribution of, 46
    stationary distribution of, 47
Markov process, 66 ff.
    ergodic, 80
    general theory of, 75
Markov property, 38, 66
Matrix
    calculating powers of, 47
    spectral resolution of, 48

Metric, in genetics, 226
Moment-generating function, 7, 12
Moments, generating function for, 7, 12
Monte Carlo methods
    in appointment systems, 147
    in prey–predator model, 193
Multidimensional process, 117 ff., 222
Multiple-phase birth process, 131
Multiplicative process, 59, 102
Mutants, 66
    in bacteria, 125
    branching process for spread of, 63
Mutation in bacteria, 125

Negative binomial distribution, 87, 99, 109
Negative exponential distribution, 68
Non-homogeneous process, 107 ff.
    chance of extinction in, 112
    effect of immigration on, 115
    for birth and death, 110
Non-Markovian processes, 222 ff.
Normal approximation to stochastic process, 208, 218
Nuclear chain reaction, 61
Null state, 44

Operations research, 136

Paramecium aurelia, 192
Partial differential equations, 69 ff
    solution of linear, 74
    solution of second-order, 202
Partial fractions, 10
    in expansions of generating functions, 10
Periodic event, 18
Periodic recurrent event, 18
Periodic state, 44
Persistent event, 18
Persistent state, 44
Poisson process, 67
    diffusion approximation for, 208
Pólya process, 107
Population growth
    birth-and-death processes for, 84 ff.
    branching processes in, 58
    diffusion processes in, 205
    with two sexes, 119

Predation, 189
  competition and, 186 ff.
Prey–predator model, 189
Probability distribution, 2
  absolute, 39
  conditional, 39
Probability process, 2
Probability-generating function, 6
Process (see under particular kind of
  process, e.g. Branching process)

Queue discipline, 137
Queue length, distribution of, 141
Queueing process, 2, 136 ff.
  continuous time treatments of, 149
  equilibrium theory of, 137
  first-passage time for, 157
  generalized, 229
  non-equilibrium treatment of, 149
  with many servers, 143
Queues, theory of, 136 ff.

Radioactive disintegration, 66, 67
Random variable, 6
"Random-variable" technique, 70
Random walk, 24
  in epidemics, 176
  one-dimensional, 25
  restricted, 24
  symmetric, 24
  three-dimensional, 35
  two-dimensional, 34
  unrestricted, 24
Recombination fraction, in genetics,
  227
Recurrence time, 17
Recurrent epidemic
  deterministic, 177
  stochastic, 179
Recurrent event, 16 ff., 38, 43
  delayed, 21, 45
Recurrent process, 223, 224
Reed and Frost's chain binomial,
  183
Reflecting barrier, 25
Regenerative process, 223
Relative removal-rate, in epidemics,
  170
Renewal process, 21, 223, 224

Renewal theory, 224
  in chromosome mapping, 226
Return to origin
  in repeated trials, 20
  in three dimensions, 36
  in two dimensions, 35
Ruin, gambler's, 25
"Rush-hour" theory, 157

Saddle-point approximations, 214
Service mechanism, of a queue, 137
Simple birth-and-death process, 91,
  232
Simple birth process, 84, 232
Simple death process, 90
Simple epidemic, 164
Spectral resolution, of a matrix, 48
Spermatozoa, movement of, 100
States of Markov chain, 39
  absorbing, 42
  aperiodic, 44
  ergodic, 44
  null, 44
  periodic, 44
  persistent, 44
  reachable, 41
  transient, 44
Stationary distribution, of Markov
  chain, 47
Stationary process, 4
Steepest descents, method of, 214
Stochastic linearization, 219
Stochastic matrix, 40
Stochastic model, 2
Stochastic process, 2
Stochastic Threshold Theorem, in epi-
  demics, 176
Struggle for existence, 186
Supplementary variables, inclusion of,
  223

Telephone calls, 66
Threshold Theorem in epidemics
  deterministic, 172
  stochastic, 176
Time series, 4
Total size of epidemic, 174
Traffic intensity, 138
Transient event, 18

Transient state, 44
Transition, 39
  matrix, 40
  probability of, 39
Transition probabilities, 39
  conditional, 77
  infinitesimal, 78

Waiting line, 136
Waiting time, 17
  in queueing theory, 141
Whittle's Stochastic Threshold Theo-
    rem, 176

Yule–Furry process, 87, 88

Applied Probability and Statistics (*Continued*)

GROSS and CLARK · Survival Distributions: Reliability Applications in the Biomedical Sciences

GROSS and HARRIS · Fundamentals of Queueing Theory

GUTTMAN, WILKS, and HUNTER · Introductory Engineering Statistics, *Second Edition*

HAHN and SHAPIRO · Statistical Models in Engineering

HALD · Statistical Tables and Formulas

HALD · Statistical Theory with Engineering Applications

HARTIGAN · Clustering Algorithms

HILDEBRAND, LAING, and ROSENTHAL · Prediction Analysis of Cross Classifications

HOEL · Elementary Statistics, *Fourth Edition*

HOLLANDER and WOLFE · Nonparametric Statistical Methods

HUANG · Regression and Econometric Methods

JAGERS · Branching Processes with Biological Applications

JESSEN · Statistical Survey Techniques

JOHNSON and KOTZ · Distributions in Statistics
Discrete Distributions
Continuous Univariate Distributions-1
Continuous Univariate Distributions-2
Continuous Multivariate Distributions

JOHNSON and KOTZ · Urn Models and Their Application: An Approach to Modern Discrete Probability Theory

JOHNSON and LEONE · Statistics and Experimental Design in Engineering and the Physical Sciences, Volumes I and II, *Second Edition*

KEENEY and RAIFFA · Decisions with Multiple Objectives

LANCASTER · An Introduction to Medical Statistics

LEAMER · Specification Searches: Ad Hoc Inference with Non-experimental Data

McNEIL · Interactive Data Analysis

MANN, SCHAFER, and SINGPURWALLA · Methods for Statistical Analysis of Reliability and Life Data

MEYER · Data Analysis for Scientists and Engineers

OTNES and ENOCHSON · Digital Time Series Analysis

PRENTER · Splines and Variational Methods

RAO and MITRA · Generalized Inverse of Matrices and Its Applications

SARD and WEINTRAUB · A Book of Splines

SEARLE · Linear Models

THOMAS · An Introduction to Applied Probability and Random Processes

WHITTLE · Optimization under Constraints

WILLIAMS · A Sampler on Sampling

WONNACOTT and WONNACOTT · Econometrics

WONNACOTT and WONNACOTT · Introductory Statistics, *Third Edition*

WONNACOTT and WONNACOTT · Introductory Statistics for Business and Economics, *Second Edition*

YOUDEN · Statistical Methods for Chemists

ZELLNER · An Introduction to Bayesian Inference in Econometrics